T0215315

Notes on
Statistics and Data Quality
for Analytical Chemists

Notes on
Statistics and Data Quality
for Analytical Chemists

Michael Thompson • Philip J Lowthian

Birkbeck University of London, UK

ICP

Imperial College Press

Published by

Imperial College Press
57 Shelton Street
Covent Garden
London WC2H 9HE

Distributed by

World Scientific Publishing Co. Pte. Ltd.
5 Toh Tuck Link, Singapore 596224
USA office: 27 Warren Street, Suite 401-402, Hackensack, NJ 07601
UK office: 57 Shelton Street, Covent Garden, London WC2H 9HE

British Library Cataloguing-in-Publication Data
A catalogue record for this book is available from the British Library.

NOTES ON STATISTICS AND DATA QUALITY FOR ANALYTICAL CHEMISTS

ISBN-13 978-1-84816-616-5
ISBN-10 1-84816-616-8
ISBN-13 978-1-84816-617-2(pbk)
ISBN-10 1-84816-617-6(pbk)

Typeset by Stallion Press
Email: enquiries@stallionpress.com

Printed in Singapore by B & Jo Enterprise Pte Ltd

Preface

This book is based on the experience of teaching statistics for many years to analytical chemists and the specific difficulties that they encounter. Many analytical chemists find statistics difficult and burdensome. That is a perception that we hope to dispel. Statistics is straightforward and it adds a fascinating extra dimension to the science of chemical measurement. In fact it is hardly possible to conceive of a measurement science such as analytical chemistry that does not have statistics as both its conceptual foundation and its everyday tool. Measurement results necessary have uncertainty and statistics shows us how to make valid inferences in the face of this uncertainty. But, over the years, it has become apparent to us that statistics is much more interesting when it makes full use of the computer revolution.

It would be hard to overstate the effect that easily-available computing has had on the practice of statistics. It is now possible to undertake, often in milliseconds and with perfect accuracy, calculations that previously would have been impracticably longwinded and error-prone. A simple example is the calculation of the probabilities associated with density functions. Moreover, we can now produce in seconds several accurate graphical images of our datasets and select the most informative. These capabilities have transformed the applicability of both standard statistical methods and more recent computer-intensive methods. Textbooks for the most part have not caught up with this revolution and, to an unnecessary degree, are still stressing pencil-and-paper methods of calculation. Of course, a small number of pencil-and-paper examples of some elementary examples can assist learning, but they are too prone to mistakes for 'real-life' application. Many textbooks place a heavy stress on the mathematical basis of statistics. We regard this as inappropriate in an applied text. Analytical chemists do not need too many details of statistical theory, so we have kept these to a minimum. Drivers don't need to know exactly how every part of a car works in order to drive competently.

With ease of computation, there is, of course, a concomitant danger that people are tempted to use one of the many excellent computer statistics packages (or perhaps one of the not-so-excellent ones) without understanding what the output means or whether an appropriate method has been used. Analytical chemists have to guard against that serious shortcoming by exercising a proper scientific attitude. There are several ways of developing that faculty in relation to statistics. A key practice is the habitual careful consideration of the data before any statistics is undertaken. A visual appraisal of a graphical image is of paramount importance here, and the book is profusely illustrated with them: almost every dataset discussed or analysed is depicted. Another essential is developing an understanding of the exact meaning of the results of statistical operations. Finally, practitioners need the experience of both guided and unsupervised consideration of many examples of relevant datasets. Drivers don't need too many details of how the car works, but they do need the Highway Code, a road . map, some driving lessons and as much practice as they can get.

The book is divided into quite short sections, each dealing with a single topic, headed by a 'Key points' box. Most sections are terminated by 'Notes and further reading' with useful references for those wishing to pursue topics in more detail. The sections are as far as possible self-contained, but are extensively cross-referenced. The book can therefore be used either in a systemic way by reading the sections sequentially, or as a quick reference by going directly to the topic of interest. Every statistical method and application covered has at least one example where the results are analysed in detail. This enables readers to emulate this analysis on their own examples. The statistical results on these examples have been cross-checked by at least two different statistics packages. All of the datasets used in examples are available for download, so that readers can compare the output of their own favourite statistical package with that shown in the book and thus verify that they are entering data correctly.

Statistics is a huge subject, and a problem with writing a book such as this is knowing where to stop. We have concentrated on providing a selection of techniques and applications that will satisfy the needs of most analytical chemists most of the time, and we make no apology for omitting any mention of the numerous other interesting methods and applications. Statisticians may be surprised at the relative emphasis placed on different topics. We have simply used heavier weighting on the topics that experience has told us that analytical chemists have most difficulty with. The book is cast in two parts, a technique-based approach followed by an application-based approach. This engenders a certain amount of overlap and duplication. Analytical chemists are thereby encouraged to create their own overview of the subject and see for themselves the relationship between tasks and techniques.

Statisticians will also notice that we use the 'p-value' approach to significance testing. This was adopted after careful consideration of the needs of analytical chemists. It greatly improves the transparency of significance testing so long as the exact meaning of the p-value is borne in mind, and we stress that meaning repeatedly. The alternative approaches tend to cause more difficulty. Experience has shown that analytical chemists find the somewhat convoluted logic of using statistical tables confusing and hard to remember. The confidence interval approach is simple but almost universally misunderstood among non-statisticians. Both of these methods also have the disadvantage of engendering the idea that the significance test can validly dichotomise reality — once you have set a level of confidence the test tells you 'yes' or 'no'. This tempts practitioners to use statistics to replace judgement rather than to assist it.

Finally, here are ten basic rules for analytical chemists undertaking a statistical analysis.

1. If you can, plan the experiment or data collection before you start the practical work. Make sure that it will have sufficient statistical power for your needs. Ensure that the data collection is randomised appropriately, so that any inferences drawn will be valid.
2. Make sure that you know how to enter the data correctly into your statistical software. After you have entered it, print it out for checking.
3. Examine the data as one or more graphical displays. This will often tell you all that you need to know. In addition it will tell you if your dataset is unlikely to conform to the statistical model that underlies the statistical test that you are proposing to use. Important features to look out for are: suspect data; lack of fit to linear calibrations; and uneven variance in regression and analysis of variance.
4. Select the correct statistical test, e.g., one-tailed or two-tailed, one sample, two sample or paired.
5. Make sure that you know exactly what the statistical output means, especially the $p-$value associated with a test of significance.
6. Be careful how you express the outcome in words. Avoid attributing probabilities to hypotheses (unless you are making a Bayesian analysis — not within the scope of this book).

7. Report the magnitude of an effect as well as its significance. Distinguish between effects that are statistically significant and those that are practically important.

8. After a regression always make plots of the residuals against the predictor variable. This will give you valuable information about lack of fit and uneven variance. It is sometimes useful to make other plots of the residuals, e.g., as a time series to detect drift in the measurement process.

9. If in doubt, ask a statistician.

10. Have fun!

Michael Thompson & Philip J Lowthian
Birkbeck University of London, UK
May 2010

*   *   *   *   *   *

Data files used in the book can be downloaded
from http://www.icpress.co.uk/chemistry/p739.html

Contents

PART 1
Statistics

Chapter 1

Preliminaries

This chapter sets the scene for statistical thinking, covering variation in measurement results and the properties of objects, and its graphical representation. The basis for statistical inference derived from analytical data is associated with the probability of obtaining the observed results under the assumption of appropriate hypotheses.

1.1 Measurement Variation

> **Key points**
> — Variation is inherent in results of measurements.
> — We must avoid excessive rounding to draw valid conclusions about the magnitude of variation.

The results of replicated measurements vary. If we measure the same thing repeatedly, we get a different result each time. For example, if we measured the proportion of sodium in a finely powdered rock, we might get results such as 2.335, 2.281, 2.327, 2.308, 2.311, 2.264, 2.299, 2.295 per cent mass fraction (%). This variation is not the outcome of carelessness, but simply caused by the uncontrolled variation in the activities that comprise the measurement, which is often a complex multistage procedure in chemical measurement. Sometimes it may appear that results of repeated measurements are identical, but this is always a false impression, brought about by a limited digit resolution available in the instruments used to make the measurement or by excessive rounding by the person recording or reporting the data. If the above data are rounded to two significant figures, they all turn into an identical 2.3%, which tells us nothing about the magnitude of the variation. Excessive rounding of data must be avoided if we want to draw valid inferences that depend on the variation (see §7.8 for guidance on rounding).

1.2 Conditions of Measurement and the Dispersion (Spread) of Results

> **Key points**
> — The dispersion (spread) of results varies with the conditions of measurement.
> — The shape of the distribution of replicated results is characteristic, with a single peak tailing away to zero on either side.
> — We must remember the difference between repeatability and reproducibility conditions.

The scale of variation in results depends on the conditions under which the measurements are repeated. If the analysis of the rock powder were repeated many times by the same analyst, with the same equipment and reagents, in the same laboratory, within a short period of time, (that is, the conditions are kept constant as far as possible) we might see the results represented in Fig. 1.2.1. These results were obtained under what is called repeatability conditions. If, in contrast, each measurement on the same rock powder is made by the same method in a different laboratory (obviously by a different analyst with different equipment and reagents and at a different time) we observe a wider dispersion of results (Fig. 1.2.2). These results were obtained in what we call reproducibility conditions. Notice

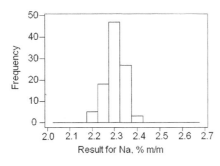

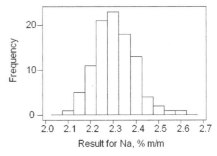

Fig. 1.2.1. Results from the analysis of a rock powder for sodium under repeatability conditions.

Fig. 1.2.2. Results from the analysis of a rock powder for sodium under reproducibility conditions.

the characteristic shape of these distributions, roughly symmetrical with a single peak tailing away to zero on either side. There are other conditions of measurement encountered by analytical chemists, but repeatability and reproducibility are the most important.

1.3 Variation in Objects

Key points
— We must distinguish between the two sources of variation (between objects and between measurement results on a single object).
— Variation among objects often gives rise to asymmetric distributions.

If we measure a quantity (such as a concentration) in many different objects in a specific category, we obtain a dispersion of results, but this is largely because the objects really do differ. Distinguishing between different objects is one of the reasons why we make measurements. Figure 1.3.1 shows the concentrations of copper measured in samples of sediment from nearly every square mile of England and Wales. As the concentrations displayed are actually the results of measurements, some of the variation (but only a small part) must derive from the measurement process. Note that this distribution is far from symmetrical — it has a strong positive skew. This skew is often observed in collections of data from natural objects.

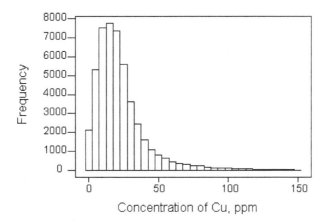

Fig. 1.3.1. Concentrations of copper in 49,300 stream sediments from England and Wales. The distribution has a strong positive skew.

1.4   Data Displays

> **Key points**
> — *Always* look at a graphical display of your data before you carry out any statistics.
> — Use visual appraisal to make a preliminary judgement about the question you are asking and to select appropriate statistical techniques.

Graphic representations of data, such as the histograms in §1.2, are essential tools in handling variation. They should always be the first resort for anyone with data to interpret. An appropriate diagram, coupled with a certain amount of experience, can tell us much of what we need to infer from data without resort to statistical methods. Indeed, a diagram can nearly always tell us which statistical techniques would suit our purpose and which of them would lead us to an incorrect conclusion.

There are several ways of representing simple replicated data. A histogram is a suitable tool to inspect the location (the central tendency) and dispersion of data all of the same type, when the number of observations is large, say 50 or more. With smaller amounts of data, histograms either look unduly ragged or do not show the shape of distribution adequately. In such circumstances the dotplot is often more helpful. (There is no exact dividing line — we have to use our own judgement!) Figure 1.4.1 shows a dotplot of the rock powder data from §1.1.

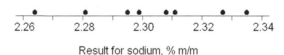

Result for sodium, % m/m

Fig. 1.4.1.   Repeated results for the concentration of sodium in a rock powder, presented as a dotplot.

1.5   Statistics

> **Key points**
> — The main reason for using statistics is the estimation of summary statistics to describe datasets in a compact way.
> — Another important reason is to assign probabilities to events and assist you in making judgements in the presence of uncertainty.

Statistics is the mathematical science of dealing with variation both in objects and in measurement results. It provides a logical way of drawing conclusions in the presence of uncertainty of measurement. It helps us in two main ways. First, it enables us to summarise data concisely, an essential step in seeing what data are telling us, especially important with large datasets. For example, the data in Fig. 1.3.1 can be summarised by very simple statistics by saying that 95% of the results fall between 15.74 and 45.36. We have condensed information about 50,000 data into three numbers. Of course this summary does not tell the whole story, but it tells us a lot more than we could find simply by looking at a list of 50,000 numbers. The histogram Fig. 1.3.1 tells us much more again but is a summary specified by about 60 numbers, namely the heights and boundaries of the histogram bars.

The other way that statistics helps us is by allowing us to assign probabilities to events. For example, it could tell us that the results obtained in an experiment were very unlikely to be obtained if certain assumptions were true. That in turn would allow us to infer that at least some of the assumptions are very likely to be untrue.

1.6 Levels of Probability

Key points

— Statistics tells us, as a probability, whether an event is likely or unlikely under stated assumptions.

— Scientists normally accept a confidence level of 95% as convincing ('statistically significant'), that is, we would observe the event with a probability of 0.95 under the assumptions, and fail to observe the event with a probability of 0.05.

— We might need a higher confidence level for certain applications.

Notice in the previous section that we are not dealing with absolutes such as 'true' or 'false', but with qualified judgements such as 'likely' or 'unlikely' etc. This uncertainty is inherent in making deductions from measurements. However, the level of probability that we accept as convincing varies both with the person making the judgement and the area of application of the result. For many scientific purposes we can accept 95% confidence. Typically we conduct an experiment to test some assumptions. Imagine that we could repeat such an experiment many times. If the results obtained in these experiments supported the assumptions only about half of the time, we wouldn't be convinced about the validity of those assumptions. If the experiment supported the assumptions 99 times out of a hundred, we would almost certainly be convinced. Somewhere between is a dividing line between 'convinced' and 'not convinced'. For many practical purposes that level is one time in twenty repeats (i.e., 95% confidence). However, we might want a much higher level of confidence under certain circumstances, in forensic science for example. To secure a prosecution based on an analytical result, such as the concentration of ethanol in a blood sample, we would want a very high level of confidence. We would not accept a situation where our result gave rise to the wrong conclusion one time in twenty. We will see later (§2.3) how these probabilities are estimated.

1.7   An Example — Ethanol in Blood — a One-Tailed Test

Key points
— If we are concerned about whether results are significantly greater than a legal or contractual limit, we calculate 'one-tailed' probabilities.
— One-tailed probabilities would apply also to other instances where there was an interest in results falling *below* such a limit.

Suppose that we had repeated measurement results such as shown in Fig. 1.7.1. A sample of blood is analysed four times by a forensic scientist and the results compared with the legal maximum limit for driving of 80 mg ethanol per 100 ml of blood. The mean of the results is above the limit, but the variation among the individual results raises the possibility that the mean is above the limit only by chance. We need to estimate how large or small that probability is. Using methods based on standard statistical assumptions to be explained in §1.8, we find that the probability of obtaining that particular mean result if the blood sample contained exactly 80 mg/100 ml is 0.0005. This low probability means that it is very unlikely that such a high mean result could be obtained if the true concentration were 80 mg/100 ml. (An even lower probability would apply if the true concentration were lower than the limit.) As the probability is very low, we infer that the sample contained a level of ethanol higher than the limit, probably high enough in this instance to support a legal prosecution.

Notice that, if we repeated the calculation for a different set of results that were closer to the limit (Fig. 1.7.2), we would obtain a higher probability estimate of 0.02. We would not, in that case, use the data to support

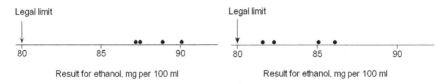

Fig. 1.7.1. Results for the determination of ethanol in a sample of blood.

Fig. 1.7.2. Results for the determination of ethanol in a different sample of blood.

a prosecution, even though the suspect was probably over the limit. The reason is that the probability of getting those results if the suspect were innocent would be unacceptably high.

Notice also that we are interested here only with probabilities of data falling above a limit: this is called a one-tailed probability. (In other circumstances we might be interested only in probabilities of data falling below a limit. An example might be testing a dietary supplement for a guaranteed minimum level of a vitamin. That would also entail calculating one-tailed probabilities.)

1.8   What Exactly Does the Probability Mean?

Key points

— The null hypothesis is an assumption that we make about the infinite number of possible results that we could obtain by replicating the measurement under unchanged conditions.

— We can calculate from the data the probability of observing the actual (or more extreme) results given the null hypothesis, but not the probability of the null hypothesis given the results.

The probabilities calculated in §1.7 have a very specific meaning. It is essential for the analyst to keep that meaning in mind when using statistics. First, it is a value calculated under a number of assumptions. A crucial assumption is that our results comprise a random sample from an infinite number of possible values. We further assume that the mean value μ of this infinite set is exactly equal to the legal limit x_L. This latter assumption is called the null hypothesis and, in this instance, is written $H_0 : \mu = x_L$.

Second, it is the probability of observing the 'event' (the results obtained or more extreme results) under these assumptions, not the probability of the null hypothesis being true given the results. In terms of the forensic example, broadly speaking it is the probability of getting the results assuming innocence, not the probability of innocence given the results. There is a crucial difference between the two probabilities. They are logically related, but can be rather different in value. The latter probability can be calculated from the former, but only if we have some extra information that cannot be derived from the data. (Using such extra information gives rise to 'Bayesian statistics', which is beyond the scope of these notes.)

1.9   Another Example — Accuracy of a Nitrogen Analyser — a Two-Tailed Test

---

Key point

— If we are concerned about whether results are significantly *different* from a reference value (such as a true value or other reference value), we need to calculate 'two-tailed' probabilities.

Suppose that we want to test the accuracy of an instrument that automatically measures the nitrogen content of food (from which we can estimate the protein content). We could do that by inserting into the instrument a sample of pure glycine, an amino acid for which we can calculate the true nitrogen content ($x_{true} = 18.92\%$), and observing the result x. Because results vary, we would probably want to repeat the experiment a number of times and compare the mean result $\bar{x}$ with x_{true}. Suppose that we obtain the results: 18.95, 18.86, 18.74, 18.93, 19.00% nitrogen (shown in Fig. 1.9.1).

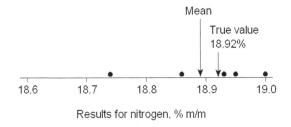

Fig. 1.9.1. Set of measurement results showing no significant difference between the true value and the mean.

The mean result is 18.90 (to four significant figures). We want to know whether the absolute difference between the true value and the mean result, $|\bar{x} - x_{true}| = |18.90 - 18.92| = 0.02\%$, plausibly represents an inaccuracy in the machine or is more probably due to the usual variation among the individual results.[1] In other words, we are asking whether $|\bar{x} - x_{true}|$ is significantly greater than zero. In this case the null hypothesis is $H_0 : \mu = 18.92$ or, more generally, $H_0 : \mu = x_{true}$. Under H_0 (and the other assumptions) we calculate the probability of getting the observed value of $|\bar{x} - x_{true}|$ or a greater one (i.e., a mean result even further from x_{true} in either direction). This particular probability has the value of $p = 0.62$. We could expect a difference as large as (or greater than) $|\bar{x} - x_{true}|$ as often as six times in ten repeat experiments if there were no inaccuracy in the instrument. As this is a large probability, greater than 0.05 for example, we say that there is no significant difference, no compelling reason to believe that the machine is inaccurate.

However, if the results had been as depicted in Fig. 1.9.2, the probability would have been $p = 0.033$, indicating a quite unusual event under the null hypothesis. We would have drawn the opposite inference, namely that there was a real bias in the results.

In these examples, in contrast to that in §1.7, we are interested in probabilities related to deviations from x_{true} in either direction, positive or negative: we want to know whether the mean is significantly *different* from the reference value x_{true}. This calls for the calculation of a two-tailed probability. (Contrast this with §1.7 where we were concerned with whether the mean was significantly *greater* than the reference value, i.e., a one-tailed probability.)

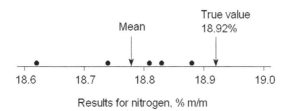

Fig. 1.9.2. Set of measurement results showing a significant difference between the true value and the mean.

---

[1] The notation $|a|$ signifies the absolute value of a, the magnitude of the number without regard to its sign, so that $|-3| = |3| = 3$.

1.10   Null Hypotheses and Alternative Hypotheses

Key points
- The alternative hypothesis for a one-tailed test is $H_A : \mu > x_L$ or $H_A : \mu < x_L$.
- The alternative hypothesis for a two-tailed test is $H_A : \mu \neq x_{true}$.

We have seen (§1.9) that calculating probabilities from results depends on the specification of a null hypothesis H_0. To distinguish between one-tailed and two-tailed situations and to allow the calculation of the correct probability we have to invoke some extra information, called the 'alternative hypothesis', which is designated H_A (or H_1 in some texts).

For a one-tailed probability the null hypothesis is $H_0 : \mu = x_L$ for a limit value x_L. For the alternative hypothesis, we would use either $H_A : \mu > x_L$ or $H_A : \mu < x_L$, depending respectively on whether x_L was an upper limit for the quantity of interest or a lower limit. For a two-tailed probability, the null hypothesis is $H_0 : \mu = x_{true}$ for a reference value x_{true}. The alternative hypothesis is then $H_A : \mu \neq x_{true}$.

The role of the alternative hypothesis is to remind us of what we are trying to establish when we are undertaking statistical calculations. It is also what we accept by default if the evidence is such as to reject the null hypothesis, that is, to find that the outcome is statistically significant.

1.11   Statements about Statistical Inferences

Key point
- We have to be very careful in our choice of words to avoid misleading statements about statistical inference.

Having settled on a level of probability that we designate as convincing for the particular inference that we wish to make (a critical level, p_{crit}), and then having estimated the probability p associated with our data under H_0 and H_A, we may wish to express the finding in words. Acceptable and unacceptable forms of words are tabulated below. They should be qualified by referencing p_{crit} in the form $100(1 - p_{crit})\%$, so if $p_{crit} = 0.05$ we would say 'at the 95% confidence level'.

$p \geq p_{crit}$	$p < p_{crit}$
We can say: We cannot reject the null hypothesis. There are no grounds for considering an alternative hypothesis. **We might say:** We accept the null hypothesis.	**We can say:** We can reject the null hypothesis and consider the alternative.
We can say [for two-tailed probabilities]: We find no significant difference between the mean result and the reference value.	**We can say [for two-tailed probabilities]:** We find a significant difference between the mean value and the reference value.
We can say [for one-tailed probabilities]: (i) We do not find the mean result to be significantly greater than the reference value; or (ii) We do not find the mean result to be significantly less than the reference value.	**We can say [for one-tailed probabilities]:** (i) We find the mean result to be significantly greater than the reference value; or (ii) We find the mean result to be significantly less than the reference value.
We cannot say: The null hypothesis is true [because statistical inference is probabilistic: there is a small chance that the null hypothesis is false].	**We cannot say:** The null hypothesis is false [because statistical inference is probabilistic: there is a small chance that the null hypothesis is true].
We cannot say: The null hypothesis is true with a probability p [because we need additional information to estimate probabilities of hypotheses].	**We cannot say:** The null hypothesis is false with a probability $(1 - p)$ [because we need additional information to make inferences about hypotheses].

We should note that it is misleading to regard 95% confidence as a kind of absolute dividing line between significance and non-significance. A confidence level of 90% would be convincing in many situations or, at least, suggest that further experimentation would be profitable.

Chapter 2

Thinking about Probabilities and Distributions

This chapter covers the estimation of probabilities relating to data from the assumption of the normal distribution of analytical error. Various approaches are covered but the main thrust is the use of the p-value to determine how likely the data are under the various assumptions.

2.1 The Properties of the Normal Curve

Key points
— Probabilities of random results falling into various regions of the normal distribution are determined by the values of μ and σ.
— To apply the normal model to estimating probabilities, we have to assume that our data comprise a random sample from the infinite population represented by the normal curve. Those may or may not be reasonable assumptions.
— Real repeated datasets encountered in analytical chemistry seldom resemble the smooth normal curve, but often provide ragged histograms.

We can estimate the probabilities involved in statistical inference in several quite different ways, but most often by reference to a mathematical model of the variation. The model most widely applicable in analytical chemistry is the normal distribution, which has a probability density (height of the curve y for a given value of x) given by

$$y = \frac{\exp((x - \mu)^2 / 2\sigma^2)}{\sqrt{2\pi}\sigma} \tag{2.1.1}$$

and a unit area. The appearance of the normal distribution depends on the values of the parameters, μ (mu) and σ (sigma). μ is called the mean of the function, and σ the standard deviation, σ^2 is called the variance. The shape of the normal curve is shown in Fig. 2.1.1. It is symmetrical about μ where the highest density lies, and falls to near-zero density at distances outside the range $\mu \pm 3\sigma$.

A key feature of a density function such as Eq. (2.1.1) is that the area defined by any two values of x represents the probability of a randomly distributed variable falling in that range. In the normal curve (Figs. 2.1.2–2.1.4), we find that about 2/3 of results fall within the range $\mu \pm \sigma$. Close to 19/20 results (or 95%) fall within the range $\mu \pm 2\sigma$, while about 99.7% fall within $\mu \pm 3\sigma$. Exact limits for 95% probability are $\mu \pm 1.96\sigma$. If we are interested in one-tailed probabilities, we note that 95% of results fall below $\mu + 1.63\sigma$ (Fig. 2.1.4) and 95% fall above $\mu - 1.63\sigma$.

Analysts should commit these ranges and probabilities to memory.

The normal curve is widely used in statistics, partly because of theory and partly because replicated results often approximate to it. The Central

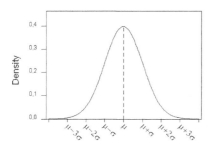

Fig. 2.1.1.   The normal curve.

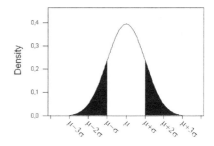

Fig. 2.1.2.   The normal curve. The region within the range $\mu \pm \sigma$ (unshaded) occupies about 2/3 of the total area.

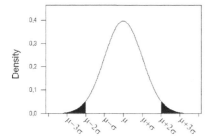

Fig. 2.1.3.   The normal curve. The region within the range $\mu \pm 2\sigma$ (unshaded) occupies about 95% of the total area.

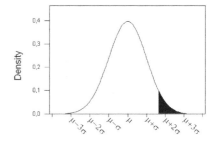

Fig. 2.1.4.   The normal curve. The region below $\mu + 1.63\sigma$ (unshaded) occupies 95% of the total area.

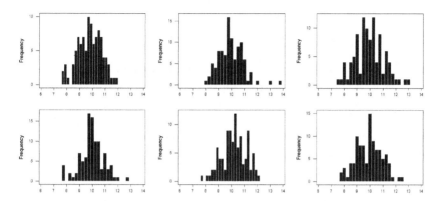

Fig. 2.1.5. Six random samples of 100 results from a normal distribution with mean 10 and standard deviation 1.

Limit Theorem shows that the combination of numerous small independent errors tends to give rise to such a curve. This combination of errors is exactly what we would expect in analytical operations, which comprise a lengthy succession of separate actions by the analyst, each prone to its own variation. In practice, repeated analytical data usually take a form that is indistinguishable from a random selection from the normal curve. Of course, histograms of real datasets are not smooth like the normal curve. Histograms are 'steppy' if representing a large dataset (e.g., Fig. 1.3.1) but tend to be ragged for the size of dataset usually encountered by analytical chemists. Genuine random selections from a normal distribution, even quite large samples, seldom closely resemble the parent curve. Figure 2.1.5 shows six such selections.

2.2 Probabilities Relating to Means of n Results

Key points

— $\sigma/\sqrt{n}$ is called the 'standard error of the mean'. 'Standard error' simply means the standard deviation of a statistic (such as a mean) as opposed to that of an individual result.

— Means of even a small number of results tend to be close to normally distributed even if the original results are not.

We are often interested in probabilities relating to the mean of two or more results. Statistical theory tells us that the mean $\bar{x}$ of n random results from the normal curve with mean μ and standard deviation σ is also normally distributed, with a mean of μ but a standard deviation of $\sigma/\sqrt{n}$. Even if the parent distribution of the individual results differs from a normal distribution, the means will be much closer to normally distributed, especially for higher n. Again, this is an outcome of the Central Limit Theorem. The term $\sigma/\sqrt{n}$ is known as 'the standard error of the mean'.

We can now apply the previously established properties of the normal curve to means. For example, the probability of an observed mean $\bar{x}$ falling above $\mu + 1.63\sigma/\sqrt{n}$ is 0.05. This is a one-tailed probability (§1.7) for an upper limit. In statistical notation[1] we have

$$\Pr\left[\bar{x} - \mu > \frac{1.63\sigma}{\sqrt{n}}\right] = \Pr\left[\frac{\bar{x} - \mu}{\sigma/\sqrt{n}} > 1.63\right] = 0.05.$$

Likewise, for a lower limit, we have

$$\Pr\left[\bar{x} - \mu < -\frac{1.63\sigma}{\sqrt{n}}\right] = \Pr\left[\frac{\bar{x} - \mu}{\sigma/\sqrt{n}} < -1.63\right] = 0.05.$$

For a probability p other than 0.05 we simply need to replace 1.63 by the appropriate coverage factor k derived from the normal distribution, to give the general formulae

$$\Pr\left[\frac{\bar{x} - \mu}{\sigma/\sqrt{n}} > k\right] = p \qquad (2.2.1)$$

and

$$\Pr\left[\frac{\bar{x} - \mu}{\sigma/\sqrt{n}} < -k\right] = p. \qquad (2.2.2)$$

For the two-tailed case, the probability of a mean falling outside the range $\mu \pm 1.96\sigma/\sqrt{n}$ is 0.05. From this we can deduce that

$$\Pr\left[\frac{|\bar{x} - \mu|}{\sigma/\sqrt{n}} > 1.96\right] = 0.05.$$

or, in general,

$$\Pr\left[\frac{|\bar{x} - \mu|}{\sigma/\sqrt{n}} > k\right] = p. \qquad (2.2.3)$$

---

[1] The notation $\Pr[\,]$ signifies the probability of whatever expression is within the square brackets.

However, we must remember that the relationship between p and k is different in one-tailed and two-tailed probabilities.

2.3 Probabilities from Data

Key points

— Equations defining normal probabilities have to be modified if you are using standard deviations s estimated from a small number (less than about 50) of data instead of the population value σ.

— The variable t (or 'Student's t') replaces the coverage factor k in these equations.

— The value of t depends on the probability required and the number of results used to calculate the mean.

To estimate probabilities from data, first we have to find which particular normal distribution (defined by μ and σ) is the best representation of our data $x_1, x_2, \ldots, x_i, \ldots, x_n$. We do that by calculating the corresponding 'sample statistics' $\bar{x}$ and s. The 'sample mean' $\bar{x}$ is the ordinary arithmetic mean given by

$$\bar{x} = \frac{1}{n} \sum_{i=1}^{n} x_i,$$

while the 'sample standard deviation' s is

$$s = \sqrt{\frac{\sum_{i=1}^{n} (x_i - \bar{x})^2}{n - 1}}.$$

The statistics $\bar{x}$ and s are estimates of the unknown parameters μ and σ, and (usually) approach them more closely as n increases. We must remember that $\bar{x}$ and s are variables, in the sense that if a series of measurements were repeated the resultant values of $\bar{x}$ and s would be different each time. As an outcome, they cannot be used directly to substitute for the parameters μ and σ in the probabilities given in Eqs. (2.2.1)–(2.2.3). Instead we have to use modified equations, in which we substitute a variable t (also called 'Student's t') for the normal coverage factor k, giving

$$\Pr\left[\bar{x} - \frac{ts}{\sqrt{n}} < \mu < \bar{x} + \frac{ts}{\sqrt{n}}\right] = 1 - p, \qquad (2.3.1)$$

from which we obtain for two-tailed probabilities

$$\Pr\left[\frac{|\bar{x} - \mu|}{s/\sqrt{n}} > t\right] = p. \tag{2.3.2}$$

In some contexts we assume a null hypothesis $H_0 : \mu = x_{ref}$ and, in such an instance, we can substitute a reference value x_{ref} for μ. This gives the corresponding expression,

$$\Pr\left[\frac{|\bar{x} - x_{ref}|}{s/\sqrt{n}} > t\right] = p. \tag{2.3.3}$$

Corresponding expressions can be obtained for one-tailed probabilities but, again, the relationship between t and p is different.

Like k, the value of t depends on the probability p. Unlike k, however, the value of t also depends on n and gets closer to k as n increases. For small n this difference is important. With $n > 50$ there is little difference: for $n = 50$ results $t = 2.01$ compared with $k = 1.96$ for a two-tailed probability of 0.05. Corresponding values of t and p are tabulated in statistics texts, and can be calculated from each other, quickly by computer but with great difficulty by hand.

2.4   Probability and Statistical Inference

Key points

— Probabilities about means of repeated results can be calculated simply by computer under H_0 and H_A.

— Care should be taken in interpreting the exact value of a probability.

— Statistics should be used to assist a decision, not to make it automatically.

We can calculate a probability associated with specific data by using Eq. (2.3.2). If we set $\frac{|\bar{x} - x_{true}|}{s/\sqrt{n}} = t$, we can obtain the value of p associated with this 'sample t-value' derived from the data under H_0 and H_A. As a two-tailed example we take the nitrogen analyser data from §1.9. We have $H_0 : \mu = 18.92$ and $H_A : \mu \neq 18.92$. Most statistics packages give the probability $p = 0.62$ directly. This is the probability of obtaining our mean value (or a value more distant from 18.92) if H_0 is true. As this is a large value (e.g., much larger than 0.05), the event would be common under repeated experiments, so there are no grounds for suspecting the truth of H_0.

These calculated probabilities are an essential guide to the assessment of experimental results. All analysis is conducted to inform a decision. Often the decision amounts to comparing some experimental results with an independent reference value of some kind — a legal or contractual limit, or a true value. The probability of our results under stated hypotheses tells us whether those assumptions are plausible or not. The decision depends on comparing our calculated probability with a predetermined 'critical' probability, typically 0.05 for general purposes. Using such critical levels is designed to help us avoid drawing unsafe conclusions but it does not relieve us of the responsibility of making the decision. A probability of say 0.07 still means that it is considerably more likely that the null hypothesis could reasonably be rejected than otherwise.

We must further avoid placing too much reliance on the exact value of the calculated probability. The calculation is based on assumptions, in particular, of randomness, independence and normality. While these are sensible assumptions, they are unlikely to be *exactly* true. This fact is especially important in the consideration of very small probabilities, which are related to the extreme tails of a distribution. Probabilities lower than 0.001 are likely to be accurate only to within an order of magnitude and are best simply regarded as 'very low', except in the hands of an expert.

Finally, we must always remember that statistics provides a method of making optimal decisions in the face of uncertainty in our data. It is not a magical way of converting uncertainty into certainty.

2.5 Pre-computer Statistics

Key points

— Before the computer age, statisticians used tabulated values of t calculated for certain fixed levels of p. To proceed with the test, the sample t value is calculated from the data and H_0. If it is *greater* than the tabulated value for the pre-selected p, the result is regarded as 'statistically significant'.

— Nowadays, most statistics packages calculate p directly from the data, H_0 and H_A. This is the simplest way of looking at probability. If p is *less* than some small pre-determined level, the outcome is unlikely under H_0 and is regarded as 'statistically significant'.

Before computers were readily available it was not practicable to calculate p from the sample t for each problem. The alternative was to use pre-calculated values of t called 'critical values' for a small number of specific probabilities. The sample t could then be compared with the tabulated value for a specific probability (usually $p = 0.05$). If the sample t exceeded the tabulated t, we would know that the probability of the event was less than 0.05, so it would be reasonable to reject H_0 at a confidence level of $100(1 - 0.05) = 95\%$. Working this out for the previous example data from §1.9, we have the statistics: $n = 5$; $\bar{x} = 18.8960$; $s = 0.1006$. These provide a sample t given by $t = \dfrac{|\bar{x} - x_{true}|}{s/\sqrt{n}} = \dfrac{|18.896 - 18.92|}{0.1006/\sqrt{5}} = 0.53$ under $H_0 : \mu = 18.92$. We need to compare this value with Student's t for a specific probability. Values of Student's t are tabulated according to the number of 'degrees of freedom'[2] which in this case equals $n - 1 = 4$. For four degrees of freedom and a two-tailed p-value of 0.05 the tabulated value of t is 2.78. Our sample value of 0.53 is well below the critical value of 2.78 for $p = 0.05$. Again this tells us that the event is not significant at the 95% confidence level.

2.6   Confidence Intervals

Key points

— $100(1 - p)\%$ confidence limits can be calculated from the data and a t value corresponding to required value of p.

— If an H_0 value falls outside this interval, we feel justified in rejecting the null hypothesis.

— Using the confidence interval approach to significance testing is mathematically related to calculating a p-value, and it provides the same answer. Calculating the actual p-value provides more information: for example, it tells you whether you are near or far from a selected critical level.

Equation (2.3.1), $\Pr\left[\bar{x} - \dfrac{ts}{\sqrt{n}} < \mu < \bar{x} + \dfrac{ts}{\sqrt{n}}\right] = 1 - p$, tells us the range in which the unknown population mean μ falls with a probability of $(1 - p)$

---

[2]The number of degrees of freedom $(n - m)$ designates the number of independent values after m statistics have been estimated from n results. If we have say ten results, the mean has nine degrees of freedom because we can calculate any one result from the other nine and the mean.

and is known as the $100(1-p)\%$ confidence interval. So if $p = 0.05$, we have a 95% confidence interval. The ends of the range are called the 'confidence limits'. The 95% confidence interval is a convenient and currently popular method of expressing the uncertainty in our measurement results. The meaning of the confidence interval is subtle, however, and widely misunderstood. Its *exact* meaning is as follows: if the experiment were repeated a large number of times, μ would fall in the calculated interval on an average of $100(1-p)\%$ occasions.

Confidence intervals provide an alternative way of assessing and visualising tests of significance. As we assume that $\mu = x_{true}$ under the null hypothesis H_0, x_{true} should fall within the calculated 95% confidence region most of the time. If it doesn't fall within the confidence interval, we feel justified in rejecting the null hypothesis. So calculating a confidence interval is one way of attributing a limiting probability to our data. All we need to do is to settle on a desired level of confidence and find the corresponding value of t from a table.

Again using the nitrogen analyser data in §1.9, we calculate the 95% confidence limits (for a two-tailed test) as $\bar{x} \pm ts/\sqrt{n} = 18.896 \pm 2.78 \times 0.1006/\sqrt{5} = (18.77, 19.02)$. The H_0 value of 18.92 falls close to the middle of this range (Fig. 2.6.1), so there are no grounds for rejecting H_0.

If we calculate 95% confidence limits for the sodium data (§1.1) and, in addition, use the information that the rock powder is a reference material

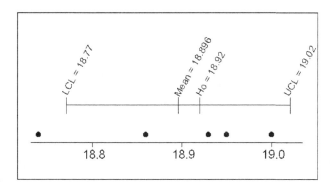

Result for nitrogen, % m/m

Fig. 2.6.1. The nitrogen analyser data (points), showing the 95% confidence interval and the H_0-value falling *inside* the interval. LCL = lower confidence limit; UCL = upper confidence limit.

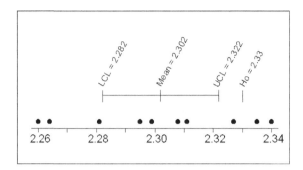

Result for sodium, % m/m

Fig. 2.6.2.   The sodium data (points), showing the 95% confidence interval and the H_0-value falling *outside* the interval. LCL = lower confidence limit; UCL = upper confidence limit.

with a certified value of 2.33% for the sodium content, we find that the certified value falls outside the confidence interval (Fig. 2.6.2), and consequently we find that there is a significant difference and justifiably consider that the analytical method is providing a biased result.

Chapter 3

Simple Tests of Significance

This chapter contains worked examples of simple tests of significance of means, incorporating one-sample tests, two-sample tests and tests on paired data, in one-tailed and two-tailed forms. Its main purpose is the demonstration of use of appropriate methods and a critical appraisal of the outcome, and to allow readers to check that they are using their own statistical computer software correctly and interpreting the output correctly. There is also a small amount of theory.

3.1 One-Sample Test — Example 1: Mercury in Fish

Key points
— This is a one-tailed test because we are concerned with concentrations *above* a regulatory limit.
— The usual symmetrical 95% confidence limits are not applicable here. We should beware that some statistical software may confusingly produce two-tailed confidence limits in combination with a one-tailed test of significance.

European Regulation 629/2008 sets a maximum concentration of mercury in fish of 1.0 ppm. A laboratory analyses a suspect sample four times and obtains the results

$$1.34, \quad 1.44, \quad 1.42, \quad 1.14 \, \text{ppm}.$$

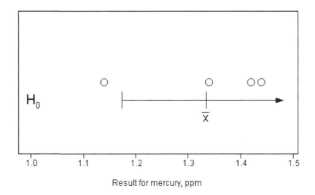

Fig. 3.1.1.   Results for mercury content with null hypothesis, mean and 95% confidence region.

Can we conclude that the concentration of mercury is above the allowable level?

This requires a one-tailed probability because we are testing the mean result against an upper limit. A display of the data is in Fig. 3.1.1: we see that the null hypothesis value does not lie in the 95% confidence region (which has only a lower limit for the one-tailed test). The statistics are shown in Box 3.1.1. The p-value of 0.0082 is low: the mean result is significantly greater than the regulatory limit. The chance of obtaining the data (or a set with a higher mean) if H_0 were true is less than one in a hundred, so we are justified in rejecting it.

Box 3.1.1 One-tailed t-test of the mean

$H_0 : \mu = 1.0 : H_A : \mu > 1.0$

Variable	n	Mean	St Dev	SE Mean	t	p
Mercury, ppm	4	1.3350	0.1370	0.0685	4.89	0.0082

Lower 95% confidence limit $= 1.17$

Note

- *The file containing this dataset is named* **Mercury**.

3.2   One-Sample Test — Example 2: Alumina (Al$_2$O$_3$) in Cement

> **Key points**
> — This is a two-tailed test because we are considering whether an observed mean differs significantly from a target value.
> — It is worth looking for trend in sequential data.

A special cement has a target for the alumina content of 4.00%. In its manufacture, the composition of the product is monitored by regular automatic analysis. Twelve successive results for alumina were:

3.91, 4.14, 3.87, 4.21, 4.05, 4.19, 4.16, 3.86, 4.13, 4.18, 4.14, 4.04.

Is there any evidence that the alumina content differs from the target 4.00% during the measurement period?

There is no obvious trend in the data (Fig. 3.2.1), so a straightforward test of significance on the mean is appropriate. The investigation calls for a two-tailed test as we are interested in a significant difference. A display of the data is shown in Fig. 3.2.2. We see that the H_0 value lies in the 95% confidence region (just).

The statistics are shown in Box 3.2.1. As the p-value is greater than 0.05, we see that the null hypothesis is not rejected at the usual 95% confidence. On the face of it, there is no compelling evidence to suggest that the true level in the sample is different from the target value (assuming

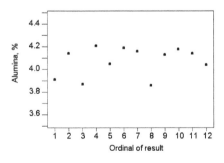

Fig. 3.2.1. The results plotted in time sequence.

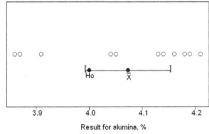

Fig. 3.2.2. Results for alumina (points), showing null hypothesis (H_0), mean and 95% confidence region.

that the analytical method is unbiased). Intervention would not be called for, because there is a small (but not negligible) probability of obtaining these results when the concentration of alumina did not differ from 4.00.

Box 3.2.1 Two-tailed test of the mean

$H_0 : \mu = 4.00 : H_A : \mu \neq 4.00$

Variable	n	Mean	St Dev	SE Mean	t	p
Alumina %	12	4.0733	0.1274	0.0368	1.99	0.071

95% Confidence limits (3.9924, 4.1543)

Notes

- *The file containing this dataset is named **Alumina**.*
- *In an instance like this one, where data have been collected sequentially, it is worth looking for a trend in the data, in case the process seems to be drifting away from the target value. That circumstance would invalidate a significance test on the mean: the data would not be independent. However, the drift itself would suggest that the process needed adjustment.*

3.3   Comparing Two Independent Datasets — Method

Key points

— If we want to test the significance of a difference between the means of two sets of results, we have to use a two-sample test.

— If we do not assume that the standard deviations of the x-data and the y-data are equal, the number of degrees of freedom has to be reduced according to a complex formula.

— Two-sample tests can provide one-tailed or two-tailed probabilities as required.

Sometimes we want to know whether the means of two separate datasets should be regarded as different. This is known as a 'two-sample' test.

Suppose that we had access to two different methods for the determination of protein nitrogen in food. The Kjeldahl method is based on the conversion of the protein nitrogen to ammonia and measuring the ammonia by titration. The Dumas method relies on the combustion of the foodstuff with the formation of elemental nitrogen which is measured volumetrically or otherwise. It is suspected that the Dumas method gives on average a slightly different result, because there are other nitrogenous substances (e.g., nitrate, nitrogenous organic bases) present in the food, which may affect the results differently. In a simple experiment we could test for this suspected bias by applying both methods repeatedly to a homogeneous sample of a foodstuff.

Suppose the Dumas results are $x_1, x_2, \ldots, x_i, \ldots, x_n$ and the Kjeldahl results are $y_1, y_2, \ldots, y_j, \ldots, y_m$. The estimated bias is $|\bar{x} - \bar{y}|$. We need to know whether $|\bar{x} - \bar{y}|$ is significantly greater than zero, or whether the measured difference is simply a result of random variations in the data. The null hypothesis for this test is $H_0 : \mu_x = \mu_y$. We set up an equation analogous to Eq. (2.3.2) by recognising that, for *any* statistic θ expected to be normally distributed, $\hat{\theta}/\text{se}(\hat{\theta})$ has a t-distribution. (The function se( ) means the standard error of whatever is in the brackets, and $\hat{\theta}$ signifies an estimate of θ.) So for the difference between the means $|\bar{x} - \bar{y}|$ we have, just like Eq. (2.3.2):

$$\Pr\left[\frac{|\bar{x} - \bar{y}|}{\text{se}(\bar{x} - \bar{y})} > t\right] = p.$$

We can readily show that $\text{se}(\bar{x} - \bar{y}) = \sqrt{s_x^2/n + s_y^2/m}$, so we have

$$\Pr\left[\frac{|\bar{x} - \bar{y}|}{\sqrt{s_x^2/n + s_y^2/m}} > t\right] = \Pr\left[|\bar{x} - \bar{y}| \times \left(s_x^2/n + s_y^2/m\right)^{-\frac{1}{2}} > t\right] = p.$$

As before, we can either use a computer to calculate p from the observed value of t, namely $\dfrac{|\bar{x} - \bar{y}|}{\sqrt{s_x^2/n + s_y^2/m}}$, or manually find whether

$\dfrac{|\bar{x} - \bar{y}|}{\sqrt{s_x^2/n + s_y^2/m}}$ exceeds the critical value of t obtained from tables for the selected level of p and the appropriate number of degrees of freedom. However, in either instance, the number of degrees of freedom for t is not $(n + m - 2)$ as might be expected: we have to make an adjustment using the complex-looking formula which gives a smaller number of degrees of freedom:

adjusted degrees of freedom

$$= \left(\frac{s_x^2}{n} + \frac{s_y^2}{m} \right)^2 \Big/ \left(\frac{s_x^4}{n^2(n-1)} + \frac{s_y^4}{m^2(m-1)} \right), \qquad (3.3.1)$$

rounded to the nearest integer.

The above is an example of a two-tailed test as we are looking for a bias between the mean results of the two methods, that is, a difference regardless of which method gives the higher result. One-tailed two-sample tests are equally possible (see §3.7).

If we can reasonably assume that the population standard deviations of the datasets, σ_x and σ_y, are equal, we can use a somewhat simpler procedure (see §3.4 below), which may be quicker for manual calculations but is of no advantage if a computer package is used.

3.4   Comparing Means of Two Datasets with Equal Variances

Key points

— If we can assume that the two datasets come from populations with the same precision, we can use the pooled standard deviation to provide a simpler version of the two-sample test. This procedure is simpler for hand-calculations.

— There is no practical advantage in using a pooled standard deviation if probabilities or *t*-values are calculated by computer.

If we can assume that the two datasets are from distributions with the same variance, the mathematics of the two-sample test of means is somewhat simpler for hand calculation. The standard error of the difference is now given by $\mathrm{se}(\bar{x} - \bar{y}) = s'\sqrt{\dfrac{1}{n} + \dfrac{1}{m}}$, where $s' = \sqrt{\dfrac{\sum_{i=1}^{n}(x_i - \bar{x})^2 + \sum_{j=1}^{m}(y_j - \bar{y})^2}{n + m - 2}}$, a standard deviation derived from both datasets, which is called the 'pooled standard deviation'. We derive probabilities from a t-value with a simple $n + m - 2$ degrees of freedom, to give

$$\Pr\left[\frac{|\bar{x} - \bar{y}|}{\mathrm{se}(\bar{x} - \bar{y})} > t\right] = p.$$

Of course the assumption that both datasets come from equal-variance populations has to be reasonable, and there is a simple method of testing that called Fisher's variance ratio test or the F-test (see §3.5).

3.5 The Variance Ratio Test or F-Test

Key point

— The F-test is used to determine whether two independent variances (or two independent standard deviations) are significantly different.

We may want to test whether two standard deviations s_x and s_y, calculated from two independent datasets $x_1, x_2, \ldots, x_i, \ldots, x_n$ and $y_1, y_2, \ldots, y_j, \ldots, y_m$, are significantly different from each other. The test statistic in this case is the ratio of the estimated variances, $F = s_x^2/s_y^2$, $s_x > s_y$, and its value depends on two separate degrees of freedom, $(n - 1)$ for x and $(m - 1)$ for y. As with t-tests, the F-test can be conducted by calculating a probability corresponding with F, or by comparing the sample value of F with tabulated values for predetermined probabilities. These probabilities depend on the assumption that the original datasets were normally distributed samples. The test can be used to determine whether it is sensible to use a pooled standard deviation for a two-sample test (§3.4) but, as will be seen, it is more widely used to determine significance in analysis of variance (§4.4).

As an example we use the nitrogen data from §3.6: the variance ratio is $F = 1.193$ and the corresponding probability under $H_0 : \sigma_x^2 = \sigma_y^2$ (against $H_A : \sigma_x^2 > \sigma_y^2$) is $p = 0.58$. In other words, if there were no difference between the variances, we would expect a ratio at least as large as 1.193 with a probability of 0.58. As this is a high probability, we show that the variances are not significantly different. The situation is illustrated in Fig. 3.5.1.

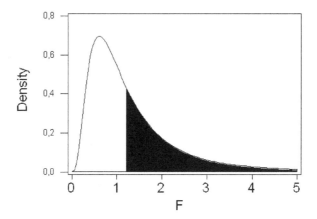

Fig. 3.5.1. The F-distribution for nine and seven degrees of freedom. The shaded area shows the probability of obtaining the nitrogen data under H_0.

Note

- *The file containing the dataset is named* **Wheat flour**.

3.6 Two-Sample Two-Tailed Test — An Example

Key points

— This is a two-tailed test because we are interested in the difference between the mean results regardless of which is the greater.

— If we assume variances are unequal when they are in fact close, we get an outcome very similar to that obtained by the pooled variance method, so no harm results.

— If we assume variances are equal when they are not we could get a misleading outcome, so we use the pooled variance method only in special circumstances.

A laboratory, possessing the two different methods (Dumas and Kjeldahl) for determining protein nitrogen in food, repeatedly analysed the same sample of wheat flour, and obtained the results given below. A dotplot of the data is shown in Fig. 3.6.1. We can see that the means of the two sets

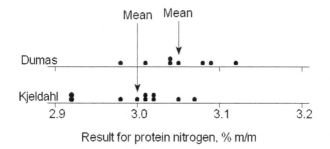

Result for protein nitrogen, % m/m

Fig. 3.6.1. Dotplot of the nitrogen data (points), showing the means of the two datasets.

of data are different, but are they significantly different, given the spread of the data? We also observe that the dispersions of the two datasets are similar.

Dumas, %	Kjeldahl, %
3.12	3.07
3.01	2.92
3.05	3.01
3.04	3.00
3.04	3.02
2.98	3.02
3.08	2.98
3.09	2.92
	3.01
	3.05

In this instance, without assuming that the variances are identical, Eq. (3.3.1) tells us to use 15 degrees of freedom and obtain $p = 0.035$ as the probability of obtaining the absolute difference that we observe (or a greater difference). We conclude that the observed difference of 0.0513% mass fraction is significant at 95% confidence.

Using a pooled standard deviation (and the full 16 degrees of freedom) provides an almost identical probability ($p = 0.036$). Notice that we do not need equal numbers of observations in each dataset for the two-sample test.

Box 3.6.1 Two-sample t-test and confidence interval

$H_0 : \mu_{Dumas} = \mu_{Kjeldahl} : H_A : \mu_{Dumas} \neq \mu_{Kjeldahl}$

Assuming unequal variances:

	n	Mean	St Dev	SE Mean
Dumas, %	8	3.0513	0.0449	0.016
Kjeldahl %	10	3.0000	0.0490	0.015

$t = 2.31 \quad p = 0.035 \quad DF = 15$
95% confidence interval for difference: $(0.004, 0.099)$

Using a pooled standard deviation:

$t = 2.29 \quad p = 0.036 \quad DF = 16$
95% confidence interval for difference: $(0.004, 0.099)$

Notes and further reading

- *The file containing this dataset is named* **Wheat flour**.
- *The outcome of this experiment may not be relevant outside the laboratory concerned. Other laboratories, conducting the same methods (but somewhat differently) may have different biases. A discussion of this point can be found in Thompson, M., Owen, L., Wilkinson, K. et al. (2002). A Comparison of the Kjeldahl and Dumas Methods for the Determination of Protein in Foods, using Data from a Proficiency Testing Scheme. Analyst, 12, pp. 1666–1668.*

3.7   Two-Sample One-Tailed Test — An Example

Key points

— This is a one-tailed test because we are interested in whether the modified conditions were better, not just different.
— Note the assumption of unequal standard deviations.

Ethanol is made industrially by the catalytic hydration of ethene. A process supervisor wants to test whether a change in the reaction conditions improves the one-pass yield of ethanol. The conversion efficiency is measured with ten successive runs under the original conditions and ten

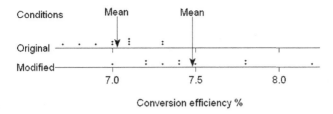

Conversion efficiency %

Fig. 3.7.1. Results for the conversion of ethene to ethanol under original and modified plant conditions.

more under the modified conditions. The results are as follows. There are no apparent trends in the data, so a simple *t*-test is appropriate.

Efficiency % original conditions	Efficiency % modified conditions
7.0	7.4
7.3	7.5
7.0	7.3
6.9	8.2
6.7	7.2
7.3	7.8
7.1	7.8
7.1	7.4
7.1	7.2
6.8	7.0

The dotplots in Fig. 3.7.1 shows the means well separated and the dispersion of the results under modified conditions greater. The *p*-value of 0.002 (Box 3.7.1) for the one-tailed test shows a very low probability of obtaining the data if the null hypothesis is true, so we can reject it and accept that the modified conditions give a significantly greater conversion efficiency. Note that by assuming unequal variances, the test uses 13 degrees of freedom in place of the original 18.

Box 3.7.1 Two-sample *t*-test and confidence interval

$H_0 : \mu_{Mod} = \mu_{Orig} : H_A : \mu_{Mod} > \mu_{Orig}$

Conditions	n	Mean	St Dev	SE Mean
Modified	10	7.480	0.358	0.11
Original	10	7.030	0.195	0.062

95% confidence interval for difference: (0.17, 0.729)
$t = 3.49$ $p = 0.0020$ DF = 13

Note

- *The file containing this dataset is named* **Ethene**.

3.8   Paired Data

Key points

— Paired data arise when there is an extra source of variation that has a common effect on both members of corresponding pairs of data.
— Paired data are treated by calculating the differences between corresponding pairs and testing the differences under $H_0 : \mu_{diff} = 0$. In this instance we have a two-tailed test so $H_A : \mu_{diff} \neq 0$.
— It is *essential* to recognise paired data: a two-sample test will usually provide a misleading answer.

Here we consider again the results of two methods for determining nitrogen in food. We want to find whether the outcome in §3.6 (i.e., no significant difference between the means) was valid for wheat in general and not just specific to a particular type of wheat. One way to do that would be to arrange for the comparison methods to be made with a number of different types of wheat. Suppose in such an experiment, in a single laboratory, the results below were obtained.

C1 Type of wheat	C2 Kjeldahl result %	C3 Dumas result %	C4 Difference %
A	2.98	3.08	0.10
B	2.81	2.88	0.07
C	2.97	3.02	0.05
D	3.15	3.17	0.02
E	3.03	3.08	0.05
F	3.05	3.21	0.16
G	3.24	3.20	−0.04
H	3.14	3.12	−0.02
I	3.04	3.11	0.07
J	3.08	3.16	0.08

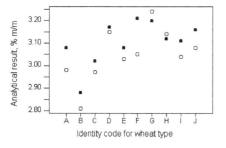

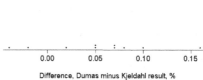

Fig. 3.8.1. Results obtained by the analysis of ten different types of wheat for the protein nitrogen content of a sample of a foodstuff. Kjeldahl result ∘. Dumas result •.

Fig. 3.8.2. Differences between corresponding results.

We must recognise that, as well as a possible difference between results of the methods, there is an extra source of variation in the results, due to variation in the true concentrations of protein in the various types of wheat. In fact both of these variations show up clearly in Fig. 3.8.1. For instance, wheat type B has provided a particularly low pair of results. It would be futile to attempt to compare the methods by comparing the mean results. If we tried to do that, any bias between methods might be swamped by the variation between the wheat types. However, we can see that it is the *differences* between respective pairs of results which will tell us what we want to know. In fact we can see in Figs. 3.8.1 and 3.8.2 that for most of the wheat types (8/10), the Dumas method gives the higher result, which in itself suggests that the difference between methods may be significant.

We deal with the situation statistically by calculating a list of differences between corresponding results. We then apply a one-sample test to the differences, with $H_0 : \mu_{diff} = 0$, i.e., we expect the mean of the differences to be zero if there is no bias between the methods. The statistics are shown in Box 3.8.1. The observed mean difference is 0.054% and we find that $p = 0.016$ (for a two-tailed test). This low value tells us that such a large observed difference is unlikely under the null hypothesis, so we feel justified in rejecting it and accepting that, for this range of wheat types, there is a bias between the methods of about 0.05%.

Box 3.8.1 Paired data two-tailed t-test

$H_0 : \mu_{diff} = 0 : H_A : \mu_{diff} \neq 0$

	N	Mean	St Dev	SE Mean
Dumas result %	10	3.1030	0.0981	0.0310
Kjeldahl result %	10	3.0490	0.1176	0.0372
Difference %	10	0.0540	0.0578	0.0183

$t = 2.96$ $p = 0.016$ DF $= 9$

95% confidence interval for mean difference: (0.0127; 0.0953)

Notes

- *The file containing this dataset is named* **Wheat types***.*
- *The bias observed in this experiment may be valid only for analyses conducted in a single laboratory. (See also the Notes in §3.6.)*

3.9   Paired Data — One-Tailed Test

Key point

— This is a one-tailed test involving paired data.

Nitrogen oxides (NO_x) are harmful atmospheric pollutants produced largely by vehicle engines. Their concentration is monitored in large cities. This experiment attempts to tell whether concentrations measured at face level are higher than those measured with a monitor at a height of 5 m (where it is safe from vandalism). One-hour average concentrations were measured every hour for a day at a particular location, with the results shown in units of parts per billion by volume (i.e., mole ratio). The results are plotted by the hour in Fig. 3.9.1. Most of the differences between methods are positive (21/24), showing a higher reading at face level. A dotplot of the differences (Fig. 3.9.2) shows the null hypothesis value well below the lower 95% confidence limit. The statistics (Box 3.9.1) show a p-value well below 0.05, so we conclude that the null hypothesis can be safely rejected in favour of the alternative: the concentration of NO_x at face level is significantly higher than at 5 m above the pavement.

Hour	NO_x, ppb at 1.5 m	NO_x, ppb at 5 m	Difference, ppb
1	11	10	1
2	15	13	2
3	16	13	3
4	13	11	2
5	19	15	4
6	16	14	2
7	15	13	2
8	20	15	5
9	18	17	1
10	19	21	−2
11	26	24	2
12	22	19	3
13	26	22	4
14	27	24	3
15	24	25	−1
16	26	22	4
17	28	24	4
18	23	25	−2
19	23	19	4
20	22	20	2
21	21	19	2
22	18	12	6
23	14	14	0
24	19	18	1

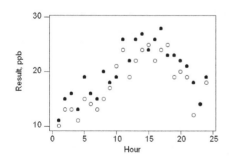

Fig. 3.9.1. Results for the concentration of NO_x in air at a site measured at two heights above ground: 1.5 m (●) and 5 m (○).

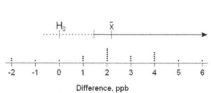

Fig. 3.9.2. Differences between pairs of results for NO_x.

Box 3.9.1 t-test of the mean difference

$H_0 : \mu_{diff} = 0 : H_A : \mu_{diff} > 0$

Variable	n	Mean	St Dev	SE Mean	t	p
Difference, ppb	24	2.167	2.036	0.416	5.21	0.0000

Lower 95% confidence boundary: 1.455

Notes

- *The file containing this dataset is named **NOx in air**.*
- *We find the difference significantly greater than zero, but in such cases we must always remember to consider the separate question of whether the difference is of important magnitude. Such a question cannot be answered without reference to an external criterion based on the use to which the data will be put. In this case we might consider a difference of about 2 ppb would be unlikely to affect decisions strongly, so could be safely ignored.*

3.10 Potential Problems with Paired Data

Key points

— For a valid test, the differences must be a random sample from a single distribution, or a reasonable approximation to that.

— That is likely to occur only if the concentrations involved are drawn from a relatively short range. If a few anomalous concentrations are present, those data can be safely deleted before the paired data test.

— It is important to recognise paired data. Treating paired data by a two-sample test will probably lead to an incorrect inference.

The paired data test is based on the hidden assumption that the differences form a coherent set. As an example, it is a reasonable assumption that the results of the Dumas method in §3.8 are taken from distributions with different means but the same (unknown) standard deviation σ_D. Likewise the Kjeldahl results are taken to represent values taken from distributions with unknown standard deviation σ_K, probably different from σ_D. These are reasonable assumptions because there is no great variation among the concentrations of nitrogen in the types of wheat. The differences then have a standard deviation of $\sigma_{diff} = \sqrt{\sigma_D^2 + \sigma_K^2}$ which we estimated for the t-test from the observed differences as $s_{diff} = 0.0578$.

If the assumption of a single standard deviation is unrealistic, the outcome of the test may be misleading. Consider the following data, which refer to the determination of beryllium in 11 different types of rock by two different analytical methods.

Rock type	ICP result Be, ppm	AAS result Be, ppm	Difference ppm
A	2.6	2.9	0.3
B	1.7	1.9	0.2
C	1.7	1.6	−0.1
D	2.4	2.5	0.1
E	0.4	1.1	0.7
F	1.2	1.5	0.3
G	1.6	1.9	0.3
H	2.9	2.9	0.0
J	2.2	2.6	0.4
K	2.1	2.4	0.3
L	56.2	61.5	5.3

If the data are treated naively by conducting a t-test under $H_0 : \mu_{diff} = 0 : H_A : \mu_{diff} \neq 0$ on the complete dataset (all 11 differences), we obtain the result $p = 0.16$ (Box 3.10.1), an apparently non-significant result that might deceive an inexperienced person. However, an inspection of the data shows that one difference is much greater than any of the others, and this is apparently owing to a much higher concentration of beryllium in Rock type 'L'. At this concentration (about 60 ppm) we would expect the standard deviation of the determination to be much greater than for the rest of the rocks (at concentrations of 3 ppm or less). Consequently, it is appropriate to delete the results pertaining to Rock 'L' from the list and then repeat the test with the remaining ten differences. This gives us the result $p = 0.0062$, a clearly significant result, contrasting sharply with the naïve result.

Analysts need have no fear that this procedure amounts to improperly 'adjusting the results'. The difference is not deleted because it is an outlier or anomalous *per se* (although it clearly is), but because Rock type 'L' obviously differs from the others by virtue of the very high concentration of beryllium present.

Box 3.10.1 Beryllium data

$H_0 : \mu_{diff} = 0 : H_A : \mu_{diff} \neq 0$

'Naïve' statistics

	n	Mean	St Dev	SE Mean	t	p
Difference, ppm	11	0.709	1.537	0.463	1.53	0.16

'Sensible' statistics

	n	Mean	St Dev	SE Mean	t	p
Difference, ppm	10	0.2500	0.2224	0.0703	3.56	0.0062

Another problem can arise when data are paired but the fact is overlooked. Treating paired datasets by a two-sample test will often eliminate any sign of a real significant difference. This happens because differences between any two paired results will often be considerably smaller than differences between sets of pairs. For instance, the paired beryllium results above, if incorrectly treated as two-sample results under the hypotheses $H_0 : \mu_{ICP} = \mu_{AAS} : H_A : \mu_{ICP} \neq \mu_{AAS}$, would give a p-value of 0.42, erroneously indicating a non-significant difference. Likewise, the paired results in §3.8 show a difference significant at 95% confidence. However, if the data are incorrectly treated as two-sample results under the hypotheses $H_0 : \mu_{Dumas} = \mu_{Kjeldahl} : H_A : \mu_{Dumas} \neq \mu_{Kjeldahl}$ we obtain an apparently non-significant p-value of 0.28, because of the relatively large differences among the nitrogen contents of the wheat types.

Pairing is not difficult to spot if we are looking out for it. The diagnostic sign of paired data is that there is extra information available about how they were collected. Often this information is explicit. In this section, for example, we have a column telling us that each pair of data is obtained from the analysis of a different rock type. However, we do not need to know the actual rock type to do the statistics. Moreover, sometimes the information about pairing is implicit or stated separately rather than as part of the dataset. In analytical data, key signs of pairing might be that the observations were made on different types of test material, or by different analysts, by different laboratories, or on different days.

Note

- *The file containing the dataset is named **Beryllium methods**.*

Chapter 4

Analysis of Variance (ANOVA) and Its Applications

This chapter treats the statistical techniques of analysis of variance (ANOVA) and its most prominent applications in analytical science. ANOVA has many applications in various sectors that utilise (rather than produce) analytical data, agricultural studies for example, but these are not covered in this book.

4.1 Introduction — The Comparison of Several Means

Key points

— Analysis of variance (ANOVA) is a broad method for analysis of data affected by two (or more) separate sources of variation.

— Typically the sources of variation are between and within subsets of results.

— Important applications of ANOVA in analytical science are in (a) homogeneity testing, (b) sampling uncertainty and (c) collaborative trials.

There are four or more recognised methods for the determination of 'fat' in foodstuffs. They are thought to give different results from each other, because 'fat' is not a clearly defined analyte (although its determination is very important commercially). Suppose that we applied the four most popular of these methods ten times each to a well-homogenised batch of dog food, and produced the results shown in Fig. 4.1.1 (Dataset A). Visually there is no apparent reason to believe that the mean results of the four methods differ significantly among themselves. We could quite readily accept the hypothesis that the four sets each comprise random selections

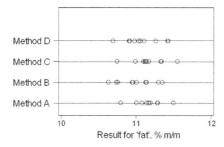

Fig. 4.1.1. Dataset A. Results from the determination of fat in dog food by using four different methods.

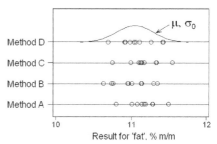

Fig. 4.1.2. Hypothesis that all of the results could be accounted for as four random samples of data from a common normal distribution.

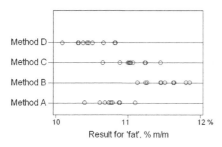

Fig. 4.1.3. Dataset B. Results from the analysis of dog food, showing differing mean values of results from four different methods.

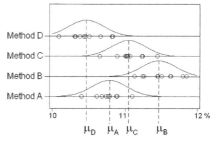

Fig. 4.1.4. Hypothesis that accounts for the results by assuming that the sets of results came from distributions with different means.

from a single normal distribution $N(\mu, \sigma_0^2)$ (Fig. 4.1.2). We could account for all of the variation by means of one mean μ and one standard deviation σ_0.

In contrast, we would entertain no such hypothesis if we obtained the results shown in Fig. 4.1.3 (Dataset B). It would seem far more plausible that the four sets of results came from four separate normal distributions, all with the same standard deviation σ_0 but with distinct means $\mu_A, \mu_B, \mu_C, \mu_D$, as in Fig. 4.1.4. In this set of data there are two separate sources of variation. There is still the common within-method variation designated by σ_0, but there is an independent variation due to the dispersion of the means. The standard deviation of the means is designated σ_1. By the principle of addition of independent variances, the variance of a single observation is $\sigma_0^2 + \sigma_1^2$, and the variance of a mean of n results from a single group is $\sigma_0^2/n + \sigma_1^2$.

The statistical method called ANOVA handles a range of problems like the one illustrated here. More complex versions of ANOVA have been devised, but only the simpler ones have regular application in analytical science. ANOVA enables us to do two things. First, it enables us to estimate separately the values of σ_0 and σ_1 from datasets such as those illustrated. This procedure has several important applications in analytical science, including: (a) testing a divided material for homogeneity (§4.5); (b) studying the uncertainty caused by sampling (§4.7); and (c) analysing the results of collaborative trials (interlaboratory studies of method performance) (§4.6).

The second range of applications of ANOVA provides us with a test of whether the differences among a number of groups of results (like the results of different methods of analysis above) are statistically significant. One way of writing the null hypothesis for this test is $H_0 : \mu_A = \mu_B = \mu_C = \cdots$. What we actually test, however, is $H_0 : \sigma_1 = 0$ versus $H_A : \sigma_1 > 0$. There are fewer important applications of this aspect of ANOVA in analytical science, partly because analysts usually need information about chemical systems that is deeper than the simple fact that there are significant differences among them.

Note

*The datasets used in this section (and §4.3) can be found in files **Dogfood dataset A** and **Dogfood dataset B**. The raw data are shown in §4.3.*

4.2   The Calculations of One-Way ANOVA

Key points

— One-way ANOVA considers a number of groups each containing several results.

— The two primary statistics calculated from the data are: (a) the within-group mean square, *MSW*; and (b) the between-group mean square, *MSB*.

— Estimates of σ_0 and σ_1 can be calculated from the mean squares.

— The null hypothesis H_0: $\sigma_1 = 0$ can be tested versus $H_A : \sigma_1 > 0$ by calculating the probability associated with the value of $F = MSB/MSW$.

One-way ANOVA is concerned with situations such as those shown in §4.1 where there is a source of variation (between groups) beyond the usual measurement errors. We consider a general case with m groups each containing n results. It is not necessary to have equal numbers of results in each group, but it simplifies the explanation and the notation. Each result x_{ji} has two subscripts, the first referring to the j-th row of data (the group) and the second to the i-th result in the row. The mean and variance of the results in the j-th row are shown as $\bar{x}_j, s_j^2$ (notice the single subscript for the row statistics). They have the usual definitions, although the notation is simplified here. $\sum_i x_i$ is used as shorthand for $\sum_{i=1}^{n} x_i$, while SS_j is shorthand for the j-th sum of squares, namely $\sum_i (x_{ji} - \bar{x}_j)^2$.

Group 1	x_{11}	x_{12}	$\cdots$	x_{1i}	$\cdots$	x_{1n}	$\bar{x}_1 = \sum_i x_{1i}/n$	$s_1^2 = SS_1/(n-1)$
Group 2	x_{21}	x_{22}	$\cdots$	x_{2i}	$\cdots$	x_{2n}	$\bar{x}_2 = \sum_i x_{2i}/n$	$s_2^2 = SS_2/(n-1)$
$\vdots$	$\vdots$	$\vdots$	$\vdots$	$\vdots$	$\vdots$	$\vdots$	$\vdots$	$\vdots$
Group j	x_{j1}	x_{j2}	$\cdots$	x_{ji}	$\cdots$	x_{jn}	$\bar{x}_j = \sum_i x_{ji}/n$	$s_j^2 = SS_j/(n-1)$
$\vdots$	$\vdots$	$\vdots$	$\vdots$	$\vdots$	$\vdots$	$\vdots$	$\vdots$	$\vdots$
Group m	x_{m1}	x_{m2}	$\cdots$	x_{mi}	$\cdots$	x_{mn}	$\bar{x}_m = \sum_i x_{mi}/n$	$s_m^2 = SS_m/(n-1)$

For one-way ANOVA we need to calculate two variances. First, we can calculate a within-group estimate by pooling the information from the m groups. This variance estimate is usually called the 'mean square within-group', designated MSW and given by

$$MSW = \frac{\sum_j SS_j}{m(n-1)}, \quad \text{which estimates } \sigma_0^2.$$

In this equation, the numerator is the total of all the group sums of squares. The denominator is the total number of degrees of freedom: as each row has $(n-1)$ degrees of freedom and there are m rows, the total is $m(n-1)$. (Note that this definition of pooled variance is consistent with that used in §3.4.)

The second estimate is called the 'mean square between-groups', designated by MSB. First, we calculate the grand mean $\bar{x}$ (with no subscript), which is the mean of the row means, $\bar{x} = \sum_j \bar{x}_j/m$. The variance of the row means is obtained by applying the ordinary formula for variance, giving $\sum_j (\bar{x}_j - \bar{x})^2/(m-1)$. This latter statistic, being based on the means of n

results, estimates $\sigma_1^2 + \sigma_0^2/n$, but is n times smaller than we want. Finally, multiplying by n, we have $MSB = \dfrac{n}{m-1} \sum_j (\bar{x}_j - \bar{x})^2$, which estimates $n\sigma_1^2 + \sigma_0^2$. From these considerations we see that $\sqrt{MSW}$ estimates σ_0, and $\sqrt{\dfrac{MSB - MSW}{n}}$ estimates σ_1.

When σ_1^2 is zero (as it would be under H_0), *but only then*, MSB also estimates σ_0^2. Thus, if H_0 is true, the expected ratio $F = MSB/MSW$ would be unity, but the value calculated from data would vary according to the F-distribution with the appropriate number of degrees of freedom (§3.5). We need to see whether the deviation of F from unity is significantly large by observing the corresponding probability.

Notes

- *The denominator in the expression estimating σ_1 is n (the number of repeat results in a group), not m (the number of groups).*
- *ANOVA calculates probabilities under the assumptions that: (a) the groups all have a common standard deviation σ_0 ; (b) errors are normally distributed; and (c) the variations within-groups and between-groups are independent. Unlike two-sample tests, ANOVA cannot readily take proper account of different variances among groups of data. Fortunately, it is often possible through good experimental design to ensure that assumption (a) is more or less correct. Where no such assurance is possible, the scientist has to use judgement about the applicability of the pooled variances and probabilities resulting from the use of ANOVA.*

4.3 Example Calculations with One-Way ANOVA

Key points

— We can use one-way ANOVA to conduct a significance test by calculating the probability associated with $F = MSB/MSW$.

— We can estimate σ_0 as $s_0 = \sqrt{MSW}$.

— Where the F-ratio is big enough, we can estimate σ_1 as $s_1 = \sqrt{(MSB - MSW)/n}$.

Suppose we apply these calculations to the two datasets used in §4.1. First we look at Dogfood dataset A (% mass fraction).

Dataset A

Method A	11.08	11.19	11.17	11.50	11.14	11.29	11.28	10.80	11.01	11.17
Method B	11.35	10.76	10.63	11.13	11.14	11.30	10.74	10.95	10.96	11.01
Method C	10.75	11.33	11.14	11.55	11.34	11.13	11.12	11.17	11.10	10.99
Method D	11.11	11.04	11.43	10.98	10.92	11.26	10.91	10.69	11.42	11.06

If we calculate the mean squares using the formulae in §4.2, we obtain $MSW = 0.048$ and $MSB = 0.063$. Under the null hypothesis $H_0 : \mu_A = \mu_B = \mu_C = \mu_D$, σ_1^2 must be zero, so both mean squares would be independent estimates of σ_0^2. We would expect the ratio $F = MSB/MSW$ to follow the F-distribution with $m - 1$ and $m(n - 1)$ degrees of freedom (see §3.5), so we can use the probability associated with F as a test of significance regarding H_0. If the value of p is low, lower than 0.05 say, the results would be unlikely to occur under H_0, so we would be justified in rejecting it.

For Dogfood dataset A we find that $F = 0.063/0.048 = 1.30$, and the corresponding probability is $p = 0.29$. With this high probability there would be no justification for rejecting H_0, so we infer that there are no significant differences among the means of the method results.

The corresponding statistics from dataset B are: $MSW = 0.048$, $MSB = 1.89$, $F = MSB/MSW = 1.89/0.048 = 39.3$, and $p < 0.0005$. With this high value of F and the small probability, the results are very unlikely to occur under the assumption of H_0, so we infer that there is a genuine difference among the means of the results of the methods, that is, $\sigma_1^2 > 0$.

Dataset B

Method A	10.68	10.79	10.77	11.10	10.74	10.89	10.88	10.40	10.61	10.77
Method B	11.85	11.26	11.13	11.63	11.64	11.80	11.24	11.45	11.46	11.51
Method C	10.65	11.23	11.04	11.45	11.24	11.03	11.02	11.07	11.00	10.89
Method D	10.51	10.44	10.83	10.38	10.32	10.66	10.31	10.09	10.82	10.46

For dataset B we can estimate σ_1 with some confidence. As MSW estimates σ_0^2, and MSB estimates $n\sigma_1^2 + \sigma_0^2$, we see that $s_0 = \sqrt{MSW}$ is the estimate of σ_0, while $s_1 = \sqrt{\dfrac{MSB - MSW}{n}}$ is the estimate of σ_1. This gives us:

$$s_0 = \sqrt{0.048} = 0.22\%;$$

$$s_1 = \sqrt{\frac{1.89 - 0.048}{10}} = 0.43\%.$$

Notes

- *The files containing the datasets in this section are named **Dogfood dataset A** and **Dogfood dataset B**.*

- *The denominator in the expression for s_1 is n (the number of results in a group) not m (the number of groups).*

- *Unless σ_1 is somewhat greater than σ_0 , estimates s_1 will tend to be very variable and not much use.*

- *In instances where the expression for s_1 provides a square root of a negative number, it is customary to set $s_1 = 0$. This simply means that σ_1 is too small to estimate meaningfully (or alternatively, that σ_0 is too large to do the job adequately).*

- *There are essentially two distinct types of application of ANOVA. The first type is where primarily we want to test for significant differences among a number (> 2) of means. The second type is where we are not interested in testing for significance (which we can often take for granted) but to estimate the separate variances σ_0^2 and σ_1^2 . It is important to be aware of this difference.*

4.4 Applications of ANOVA: Example 1 — Catalysts for the Kjeldahl Method

---

Key points

— In this fixed effect experiment, there were predetermined categorical differences between the groups of results.

— As such the experiment left unanswered some important, more general, questions about the methods.

In this experiment we consider the effects of different catalysts in the results of the Kjeldahl method for determining the protein content of a meat product. It is hoped that the commonly-used catalysts HgO and SeO_2, which leave toxic residues that are difficult to dispose of, can be replaced by the copper-containing catalysts that do not have that problem. The determination is carried out by one analyst, ten times with each catalyst, keeping all of the conditions as similar as possible. The results and a dotplot are shown below (Fig. 4.4.1). The groups have similar dispersions so the assumption of a common group variance (see §4.2) is reasonable, although there is a low suspect result among those for CuO/TiO_2. Variation among

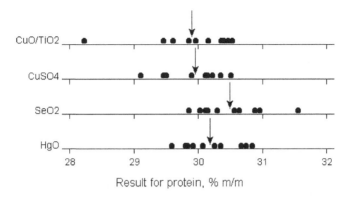

Fig. 4.4.1. Results (points) for different catalysts showing variation among means (arrows).

the group means is visible but, given the dispersion of the results, it is not clear whether the variation is significant.

HgO	SeO$_2$	CuSO$_4$	CuO/TiO$_2$
29.93	30.12	29.47	30.47
30.68	30.89	30.51	30.36
29.59	30.64	29.90	29.47
29.81	30.57	30.14	30.41
30.75	31.56	29.51	30.16
30.85	30.30	30.22	29.61
30.26	30.15	30.35	30.54
29.84	30.96	29.11	29.97
30.35	30.03	30.12	29.86
30.07	29.86	30.15	28.23

ANOVA gives the following results.

Box 4.4.1

Source of variation	Degrees of freedom	Sum of squares	Mean square	F	p
Between group	3	2.314	0.771	2.70	0.060
Within group	36	10.293	0.286		
Total	39	12.607			

The probability is $p = 0.06$, which is low enough to make us suspect that there is a real difference among the mean results of the catalysts, even though the confidence does not quite reach the 95% level.

Notes

- *The dataset used in this section can be found in the file **Kjeldahl catalysts**.*
- *There are a number of practical problems with this experiment. First, we do not know what would happen if different types of foodstuffs were used instead of the meat product: the catalysts might behave differently with other types of food, or at other concentrations of protein, or simply under the different conditions found in other laboratories. Second, we do not know enough about how the analyses were done. There might be systematic effects due to changed conditions if the results for each catalyst were obtained in a temporal sequence, especially if they were done of different days. The only way to avoid that would be to do the determinations in a randomised sequence (which could be difficult to organise in practice!).*
- *The above is an example of a **fixed-effect experiment**, which is conducted when we wish to see whether there is a significant effect when **categorical differences** between the groups have been deliberately brought about by the experimenter. There are relatively few instances in analytical science where this type of experiment is conducted. More frequently (§4.5– 4.8) we encounter experiments with **random effects** between the groups.*

4.5 Applications of ANOVA: Example 2 — Homogeneity Testing

Key points

— A simple test based on ANOVA permits us to test for lack of heterogeneity in a material.

— The usefulness of such a test depends critically on the precision of the analytical method.

A commonly encountered application of ANOVA in analytical chemistry is where we are testing a material (usually some kind of reference material) for homogeneity, i.e., to make sure that there is no measurable difference between different portions of the material. ('Measurable' here signifies under appropriate test conditions — we can usually find some difference given sufficient resources.) This is a very common exercise carried out by proficiency test providers and laboratories that manufacture reference materials. The bulk material is carefully homogenised (ground to a fine powder if solid and thoroughly mixed). It is then divided and packed into the portions that are to be distributed. A number of the packaged portions (10–20) are selected at random and analysed in duplicate (with the analysis being carried out in a random order). The generalised scheme is as shown in Fig. 4.5.1.

In this example, silicon as SiO_2 is determined in a rock powder. The results are:

Random portion	Result 1, %	Result 2, %
1	71.04	71.17
2	71.05	71.04
3	70.91	71.09
4	70.98	71.04
5	70.91	70.99
6	71.11	71.06
7	70.93	70.96
8	71.13	71.15
9	71.06	71.01
10	71.04	71.09

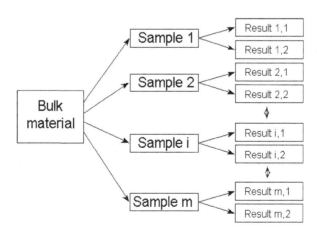

Fig. 4.5.1.   General experimental layout in a homogeneity test, with $m > 9$.

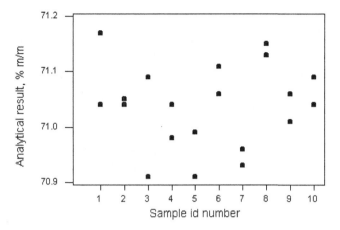

Fig. 4.5.2. Result of duplicate determination of SiO_2 in ten random samples from a bulk powdered rock.

Figure 4.5.2 shows that there are no 'outlying' sample sets and no unusually large differences between corresponding duplicate results. (At first glance we might suspect the large difference in the sample-3 results, but that is only slightly greater than that of sample 1, which in turn is only slightly greater than that of sample 5, so there is a gradation of differences rather than a single outstanding difference.) So we don't see any suspect results or obvious significant differences among the mean results. That, of course, is the expected outcome, as the material has been carefully homogenised — the test is simply to make sure that nothing has gone wrong. We now carry out the ANOVA and obtain the following results.

Box 4.5.1

Source of variation	Degrees of freedom	Sum of squares	Mean square	F	p
Between samples	9	0.07302	0.00811	2.38	0.097
Between analyses	10	0.03410	0.00341		
Total	19	0.10712			

Here we see that $p = 0.097 > 0.05$, so we cannot reject H_0 at the 95% level of confidence. As expected we find no significant differences among the ten mean results and hence the material passes the homogeneity test.

More strictly, we should say that no significant heterogeneity was found by using that particular analytical method. Use of a more precise analytical method, resulting in a smaller mean square MSB, would enhance the F-value and could easily provide a significant outcome for the same material. In practice, a more sophisticated test for lack of homogeneity, based on the same analysis of variance, is preferable to a simple test of $H_0 : \sigma_1^2 = 0$. The reason is that the test above compares the between-sample variation with the variation of the analytical method used in the test. This latter variation could be irrelevant to the users of the material. It is better to compare σ_1 with a criterion based on user needs. That, however, is beyond the scope of the current text.

Notes and further reading

- *The dataset used in this section can be found in the file named* **Silica**.
- *This example demonstrates one-way ANOVA with 'random effects'. There are two sources of random variation acting independently, namely variation between the true concentrations of silica in the samples and variation between the replicated results on each sample. This contrasts with 'fixed effects' (§4.4) where the variation between groups is imposed by the experimenter.*
- *Fearn, T. and Thompson, M. (2001). A New Test for Sufficient Homogeneity. Analyst,* **126**, *pp. 1414-1417.*
- *'Test for sufficient homogeneity in a reference material'. (2008). AMC Technical Brief No. 17A. (Free download via www.rsc.org/amc.).*

4.6   ANOVA Application 3 — The Collaborative Trial

Key point

— A collaborative trial (interlaboratory method performance study) allows us to estimate the repeatability standard deviation σ_r and the reproducibility (between laboratory) standard deviation σ_R.

The collaborative trial is an interlaboratory study to determine the characteristics of an analytical method. The main characteristics determined are **the repeatability** (average within-laboratory) standard deviation and **the reproducibility** (between-laboratory) standard deviation. These are

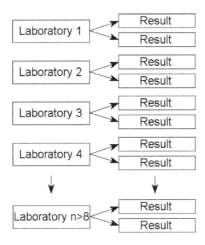

Fig. 4.6.1.   Experimental layout for one material in a collaborative trial.

obtained from the results of an experiment where each participant laboratory (of at least eight) analyses a number of materials (at least five) in duplicate by a method prescribed in considerable detail (i.e., they all use the same method as far as possible). The layout of the experiment for each material is shown in Fig. 4.6.1.

The results for each material are subjected separately to one-way ANOVA. A typical set of results (ppm), for the concentration of copper in one type of sheep feed, is as follows.

Lab ID	Result 1, ppm	Result 2, ppm
1	2.0	3.6
2	1.4	1.7
3	2.2	2.3
4	2.6	2.7
5	2.8	3.0
6	1.3	2.4
7	2.1	2.7
8	1.7	1.3
9	3.7	3.3
10	2.4	2.2
11	1.4	2.3

We can see in Fig. 4.6.2 that the variation among laboratories is roughly comparable with that between the duplicate results of any one laboratory. There are no clearly outlying laboratories or suspect duplicate

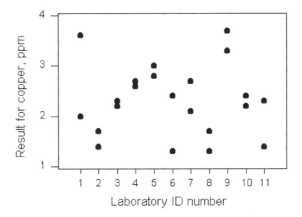

Fig. 4.6.2.   Results of the duplicate analysis of a batch of sheep feed for copper in 11 different laboratories.

differences, so we accept the data as it is. Laboratory 1 has produced the biggest differences between results, but it is not much bigger than those of Laboratory 6 or 11. (In the collaborative trial outliers, up to a certain proportion according to a strict protocol, are conventionally deleted, because we are interested in the properties of the method, not the behaviour of the laboratories — see §9.7.) ANOVA gives us:

Box 4.6.1

Source of variation	Degrees of freedom	Sum of squares	Mean square	F	p
Between laboratories	10	7.574	0.757	3.06	0.040
Within laboratories	11	2.725	0.248		
Total	21	10.299			

We calculate:

$$s_0 = \sqrt{MSW} = \sqrt{0.248} = 0.50 \, \text{ppm}$$

$$s_1 = \sqrt{\frac{MSB - MSW}{n}} = \sqrt{\frac{0.757 - 0.248}{2}} = 0.504 \, \text{ppm}.$$

The 'repeatability standard deviation' is $s_r = s_0 = 0.50$.

The 'reproducibility standard deviation' is defined as

$$s_R = \sqrt{s_0^2 + s_1^2} = 0.71 \,\text{ppm}.$$

Notes

- *The dataset used in this section can be found in the file **Copper**.*
- *Notice that $n = 2$ as we have duplicate determinations (see §4.3).*
- *There is more information about collaborative trials in §9.7 and 9.8.*

4.7 ANOVA Application 4 — Sampling and Analytical Variance

Key points

— Sampling almost always precedes analysis.

— Sampling introduces error into the final result. Because sampling targets are heterogeneous, samples from the same target differ in composition.

— ANOVA, applied to the results from a properly designed experiment, can give useful estimates of the sampling standard deviation.

Another type of application involving sampling is where we want to quantify the variance associated with sampling. Nearly all analysis requires the taking of a sample, a procedure that itself introduces uncertainty into the final result. Suppose that a routine procedure calls for the taking of a sample of soil from a field by a carefully described method and the analysis of the sample by another carefully described analytical method. We can design a suitable experiment to estimate the separate variances associated with sampling and analysis. We take a number of samples from the field by the given procedure, but randomise the method each time. We then analyse the samples in duplicate in a completely random order. This looks like a similar experiment to that in §4.5, and the schematic for the experiment (Fig. 4.7.1) is the same, but there is a difference: here we expect the samples to vary in composition, because soil is often very heterogeneous. So we are not really interested in a significance test — we can assume from the start that the samples are different but we want to know by how much they differ.

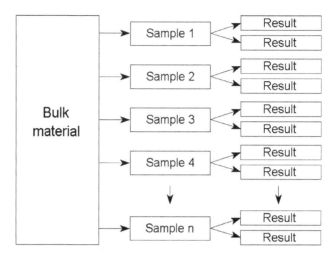

Fig. 4.7.1.   A simple design for the determination of sampling variance and analytical variance.

For an example we use the following results for cadmium (ppm). (Note: cadmium levels are exceptionally high in this field.)

Soil sample	Result 1	Result 2
1	11.8	9.8
2	6.4	6.3
3	11.9	10.3
4	12.2	10.2
5	7.5	7.3
6	6.4	6.4
7	10.1	10.0
8	11.3	9.9
9	14.0	12.5
10	16.5	15.1

Figure 4.7.2 shows (as to be expected) considerable differences between the samples and rather less between the duplicate results on each sample. There are no suspect data (i.e., seriously discrepant duplicate results on a sample, or wildly outlying samples — as judged from the mean result of the duplicates). There is no obvious reason to doubt the usual assumptions.

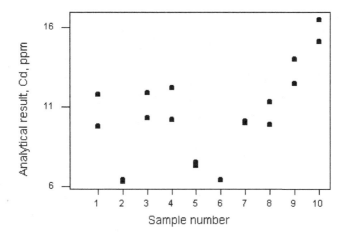

Fig. 4.7.2. Duplicate results from the analysis of ten samples of soil from a field. (For Sample 6 the results coincide.)

One-way ANOVA gives the following.

Box 4.7.1

Source of variation	Degrees of freedom	Sum of squares	Mean square	F	p
Between sample	9	160.055	17.784	21.18	0.000
Between analysis	10	8.395	0.840		
Total	19	168.449			

From the mean squares we calculate the analytical standard deviation as

$$s_a = s_0 = \sqrt{MSW} = \sqrt{0.84} = 0.92 \, \text{ppm}.$$

The sampling standard deviation is given by:

$$s_s = s_1 = \sqrt{\frac{MSB - MSW}{n}} = \sqrt{\frac{17.78 - 0.84}{2}} = 2.9 \, \text{ppm}.$$

(Notice that $n = 2$ because we have duplicate analysis — see §4.3.)

Here we see, as often happens, that the sampling standard deviation is substantially greater than the analytical standard deviation. We can see whether these variances are 'balanced'. The total standard deviation for a combined operation of sampling and analysis is going to be

$$s_{tot} = \sqrt{s_s^2 + s_a^2} = \sqrt{2.91^2 + 0.92^2} = 3.05.$$

The analytical variation makes hardly any contribution to this total variation — the sampling variation dominates. If the analytical standard deviation were much smaller at 0.46 (e.g., the analytical method was twice as precise as it is), the total standard deviation would be

$$\sqrt{2.91^2 + 0.46^2} = 2.94,$$

that is, hardly changed. There is no point trying to improve the precision of the analytical method, because it will cost much more money with no effective improvement in uncertainty of the overall result. As a rule of thumb analysts should try to get

$$1/3 < \sigma_a/\sigma_s < 3.$$

Notes

- *The dataset used in this section can be found in the file* **Cadmium.**
- *There is more information about sampling variance in Chapter 12.*

4.8   More Elaborate ANOVA — Nested Designs

Key points
— Nested designs are used in combination with ANOVA when there are two or more sources of measurement error.
— They are typically used by analytical chemists for orientation surveys (method validation) and studying uncertainty from sampling.

Hierarchical (or 'nested') designs accommodate datasets that have more than one independent source of error beyond the simple measurement error. They have a range of applications in analytical science. In an example like Fig. 4.8.1, we see that multiple fields have been sampled in duplicate, and each sample has also been analysed in duplicate. From the results of such an experiment, we can estimate three variances: the analytical variance $\sigma_a^2(= \sigma_0^2)$, the sampling variance $\sigma_s^2(= \sigma_1^2)$ and the between-site variance

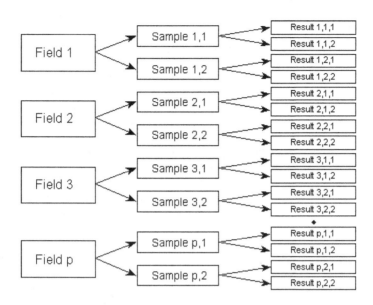

Fig. 4.8.1. Balanced nested design for an 'orientation survey'.

$\sigma_{site}^2 (= \sigma_2^2)$. In such an example, all three levels of variation can be regarded as random.

This is a classic design (known as an 'orientation survey') for validating a measurement procedure comprising sampling plus analysis, and giving some guidance to its fitness for purpose. For example, if the complete measurement variance $\left(\sigma_a^2 + \sigma_s^2\right)$ is considerably smaller than σ_{site}^2, there is a reasonable chance of differentiating between different sites by sampling and analysis. The design also provides a more rugged estimate of the sampling variance σ_s^2 than the design shown in §4.7, and is favoured for the study of sampling uncertainty. This is because the estimate will be averaged over a number of typical sites, rather than just one site, and so will be more representative of sites in general.

As an example we consider the capabilities of protocols for analysis and sampling proposed for a major survey of lead in playground dust. Ten playgrounds were selected in an inner city area and were sampled in duplicate. Each sample was then analysed in duplicate. The results were as follows (ppm of lead in dried dust). In the column headings, S1A1 indicates the first analytical result on the first sample, S1A2 the second analytical result on the first sample and so on. (These very high results refer to a period before leaded petrol was banned.)

Playground	S1A1	S1A2	S2A1	S2A2
1	714	719	644	602
2	414	387	482	499
3	404	357	380	408
4	759	767	711	636
5	833	777	748	788
6	621	602	532	520
7	455	472	498	389
8	635	708	694	684
9	589	609	591	606
10	783	764	857	803

Box 4.8.1 Analysis of variance for results for lead

Source of variation	Degrees of freedom	Sum of squares	Mean square	F	p
Between sites	9	764525	$MS_2 = 84947$	22.58	0.000
Between samples	10	37613	$MS_1 = 3761$	3.94	0.004
Analytical	20	19115	$MS_0 = 956$		
Total	39	821253			

From the mean squares we can calculate the following.

$$\text{Analytical standard deviation} = \sqrt{MS_0} = \sqrt{956} = 31\,\text{ppm.}$$

$$\text{Between-sample standard deviation} = \sqrt{\frac{MS_1 - MS_0}{2}}$$

$$= \sqrt{\frac{3761 - 956}{2}} = 37\,\text{ppm.}$$

$$\text{Between-site standard deviation} = \sqrt{\frac{MS_2 - MS_1}{4}}$$

$$= \sqrt{\frac{84947 - 3761}{4}} = 143\,\text{ppm.}$$

In this instance the proposed methods would be able to differentiate reasonably well between different playgrounds of the type in the survey. This can be seen in a plot of the results (Fig. 4.8.2).

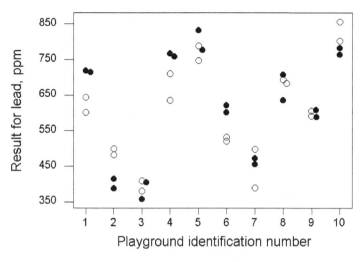

Fig. 4.8.2. Results from duplicated sampling and analysis of ten playgrounds, distinguishing between results from first sample (solid circles) and second sample (open circles).

Notes

- *The dataset used in this section can be found in the file **Playground**.*
- *There is another example of a nested experiment in §12.6.*

4.9 Two-Way ANOVA — Crossed Designs

Key points

— Crossed designs are used when we want to study results classified by two factors.

— There are relatively few applications of this technique in analytical chemistry itself, but numerous examples using analytical data in various application sectors.

Crossed designs are seldom used in analytical science as such, but the following experiment serves as an example. An investigation into the loss of weight on drying of a foodstuff subjected to different temperatures and times of heating provided the data below (% loss), illustrated in Fig. 4.9.1. There are two imposed sources of variation plus the random measurement variation. Inspection suggests that there is little difference in weight loss between one and three hours at any temperature, but that at 15 hours

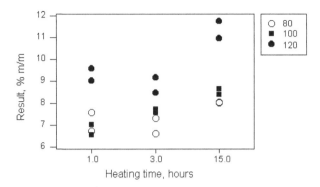

Fig. 4.9.1. Results showing weight loss at different temperatures and durations of heating.

heating there is an extra weight loss at all three temperatures. Likewise, there is little difference between corresponding results at 80° and 100° C, but the loss is more marked at 120° C.

	1 hour	3 hours	15 hours
80° C	7.57	6.61	8.02
	6.74	7.32	8.05
100° C	7.03	7.72	8.40
	6.55	7.59	8.65
120° C	9.04	8.48	10.99
	9.60	9.20	11.76

Two-way crossed ANOVA gives the following results.

Box 4.9.1 Analysis of variance of the moisture data

Source of variation	Degrees of freedom	Sum of squares	Mean square	F	p
Time	2	9.3050	4.6525	28.60	0.000
Temp	2	21.8284	10.9142	67.08	0.000
Time*Temp	4	2.2619	0.5655	3.48	0.056
Error	9	1.4643	0.1627		
Total	17	34.8596			

This output is telling us that both temperature and time separately have significant effects on the result of the drying, with p-values well below 0.05. Moreover, as we have duplicated results in all cells of the table, ANOVA can provide an estimate of the interaction between temperature and time (Time*Temp in the table) and in fact finds that this interaction should probably be regarded as significant with $p = 0.056$ (although the level of confidence does not quite reach 95%). We can see that interaction term is largely due to the fact that the loss of weight at 15 hours for 120° C is greater than would be predicted by adding the individual effects of time and temperature. While providing a test of significance, the ANOVA itself does little to help us — we have to use the diagram to see what it means.

Note

- *The dataset used in this section can be found in the file **Moisture**.*

4.10 Cochran Test

Key points

— The Cochran test compares the variances for a number of datasets. Its purpose is to determine if the largest variance is significantly larger than the others.
— It is primarily used to test for uniformity before analysis of variance.

The Cochran test compares the highest individual variance with the sum of the variance for all the datasets. If each data set contains only two values then the standard deviation (s) is replaced by the range (d) in the equation below. The test statistic is calculated as:

$$C = s_{\max}^2 \bigg/ \sum_{i=1}^{m} s_i^2$$

where m is the number of groups. If the calculated value exceeds the critical value the largest variance is considered to be inconsistent with the variances of the other datasets. There are two parameters in this test, the number of groups (m) and the degrees of freedom ($\nu = n - 1$).

An example using the Kjeldahl catalysts data from §4.4 gives the following variances for the four catalysts.

HgO	0.1915
SeO$_2$	0.2721
CuSO$_4$	0.1982
CuO/TiO$_2$	0.4820

The test statistic is C $= 0.4820/1.1438 = 0.4214$. In this example $m = 4$ and $n = 10$ (so $\nu = 9$). The critical value from tables is 0.5017 so as 0.4214 is less than this value, the null hypothesis is retained. The variance for catalyst CuO/TiO$_2$ is not regarded as inconsistent with the other variances.

Notes

- *Tables of critical values for this test are available both in textbooks and online.*
- *Use of the Cochran test in collaborative trials can be found in §9.7 and 9.8.*

4.11 Ruggedness Tests

Ruggedness is the capacity of a method to provide accurate results in the wake of minor variations in procedure such as might occur when the procedure is undertaken in different laboratories. Subjecting an analytical method to a collaborative (interlaboratory) trial (§4.6) is costly (around £50,000 at 2010 prices), so it is important that the methods tested have no unexpected defects. A ruggedness test comprises a relatively inexpensive means of screening a method for such defects.

An analytical method is made up of a moderate number of separate steps, carried out sequentially, each step carefully defined. However, analysts are expected to use some judgement in the execution of a prescribed method. For instance, if the procedure says 'boil for one hour', most analysts would expect the method to be equally accurate if the actual period was between 55 and 65 minutes, and act accordingly. Ruggedness can be tested by subjecting each separate step in the method to plausible levels of such variation and observing the measurement results. As there are many stages, this may take some time. A very economical alternative

Table 4.11.1. Experimental design for a ruggedness test.

	Experiment number							
	1	2	3	4	5	6	7	8
Factor 1	1	1	1	1	−1	−1	−1	−1
Factor 2	1	1	−1	−1	1	1	−1	−1
Factor 3	1	−1	1	−1	1	−1	1	−1
Factor 4	1	1	−1	−1	−1	−1	1	1
Factor 5	1	−1	1	−1	−1	1	−1	1
Factor 6	1	−1	−1	1	1	−1	−1	1
Factor 7	1	−1	−1	1	−1	1	1	−1
Result	x_1	x_2	x_3	x_4	x_5	x_6	x_7	x_8

is to test variations in all of the steps simultaneously. A special design for such an experiment has been developed by Youden.

The Youden design requires 2^n experiments for testing up to $2^n - 1$ factors (i.e., the steps of the analytical procedure that are under test). A widely useful size for analytical chemistry is eight experiments, which can test up to seven factors. Each experiment comprises the method with a particular combination of the factors at perturbed levels. In Table 4.11.1, the lower perturbed levels are indicated by −1 and the higher levels by 1. For instance, if the method protocol said 'boil for one hour', the respective perturbed levels tested might be 50 and 70 minutes. The combinations shown are a special subset of the 128 possible different combinations. The result of each experiment is the apparent concentration of the analyte. If the original method were completely rugged (completely insensitive to the combinations of changes) the variation in the results would estimate the repeatability standard deviation of the method. Given an appropriate choice of perturbed levels and minor effects, the variation might be somewhat larger.

The effect of a factor is estimated by the mean result for the higher level minus the mean result for the lower (or vice versa — it doesn't matter). So for (say) Factor 2 the effect is

$$\frac{x_1 + x_2 + x_5 + x_6}{4} - \frac{x_3 + x_4 + x_7 + x_8}{4}.$$

The results for the higher level group and the lower level group contain the effects of each of the other six factors exactly twice, so the extraneous effects 'cancel out'. The design ensures that this cancelling occurs with every factor.

The effects of all the factors should be listed and compared. If there are no significant effects, they will all be of comparable magnitude. A significant effect would be much greater than the majority. The method is most effective if there are no significant effects or only one. That would be the expected outcome for a properly developed method. A minor ambiguity of interpretation would occur if there were an interaction between two (or more) of the factors. For example, an interaction between Factors 6 and 7 could be mistaken for (or masked by) a main effect due to Factor 1. Such interactions should be rare in a null test (i.e., with no main effects) or a test with only one significant main effect.

Example

The analysis under consideration is the determination of patulin (a natural contaminant) in apple juice. The critical factors in the analytical method and their perturbed levels are shown in Table 4.11.2. As there are six factors we need an eight-determination layout. The seventh factor (labelled 'Dummy') needed to bring the number of factors to the nominal seven

Table 4.11.2.   Result of the ruggedness test applied to a method for the determination of patulin.

Procedure, quantity, unit	Experiment number							
	1	2	3	4	5	6	7	8
1. Extract with ethyl acetate, volume (ml)	12	12	12	12	8	8	8	8
2. Clean-up with Na_2CO_3 solution, concentration $(g/100\,ml)$	1.8	1.8	1.4	1.4	1.8	1.8	1.4	1.4
3. Clean-up duration (s)	80	40	80	40	80	40	80	40
4. Evaporate solvent, final temperature ($^\circ$C)	50	50	30	30	30	30	50	50
5. Dissolve residue in dilute acetic acid, volume (ml)	1.05	0.95	1.05	0.95	0.95	1.05	0.95	1.05
6. Determination by HPLC, injection volume (μl)	100	80	80	100	100	80	80	100
7. (Dummy)								
Analytical result, ppm	95.8	93.6	101.4	95.4	77.8	78.4	85	91.4

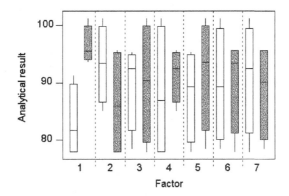

Fig. 4.11.1. Paired boxplots of the results for each factor (as numbered in Table 4.11.2). The unshaded boxes represent results for the lower perturbed levels, the shaded boxes the upper perturbed levels.

can be regarded as the null operation, i.e., 'Do nothing'. The perturbed levels were arranged as in Table 4.11.1, showing the conditions for the eight experiments and the corresponding analytical results.

We can make a good appraisal of this outcome simply by observing the results as boxplots paired for each factor (Fig. 4.11.1). We immediately see that Factor 1 (volume of ethyl acetate) seems to indicate a significant effect, Factor 2 (concentration of the clean-up reagent) a possible effect and the other factors no effect. The dummy factor (No. 7) gives a useful simulation of a null effect.

We can also calculate the effects. As an example we consider the effect of the Factor 7, that is:

$$\frac{95.8 + 93.6 + 101.4 + 95.4}{4} - \frac{77.8 + 78.4 + 85.0 + 91.4}{4} = 13.4 \, \text{ppm}.$$

A complete list of factors and effects, in decreasing magnitude, is as follows.

Procedure	Effect, ppm
1. Extract with ethyl acetate, volume (ml)	13.4
2. Clean-up with Na_2CO_3 solution, concentration (g/100 ml)	−6.9
5. Dissolve residue in dilute acetic acid, volume (ml)	3.8
4. Evaporate solvent, final temperature (°C)	3.2
7. (Dummy)	−2.4
6. Determination by HPLC, injection volume (μl)	0.5
3. Clean-up duration (s)	0.3

The repeatability standard deviation of the method was separately determined to be 8 ppm at this concentration, so the expected standard deviation of a null effect should be about 5.7 ppm. This shows that the volume of ethyl acetate probably has a significant effect and should be more carefully controlled. None of the other factors are significant on that basis. As there is only one significant effect, the possibility of confounding interactions can be ignored in this context.

Further reading

• *Youden, W.J. and, Steiner, E.H. (1975). Statistical Manual of the AOAC, AOAC International, Washington DC. ISBN 0-935584-15-3.*

Chapter 5

Regression and Calibration

Linear regression is the natural approach to analytical calibration: it is a method of fitting a line of best fit (in some defined sense) to calibration data. Regression is capable of telling us nearly all that we need to know about the quality of a well-designed calibration dataset, including the likely uncertainty in unknown concentrations estimated thereby. It also has an application in comparing the results of different analytical methods.

5.1 Regression

Key points

— Regression is a method for fitting a line to experimental points.
— Regression uses experimental points x_i, y_i, $(i = 1, \ldots, n)$, and these points are taken as fixed numbers when the experiment is complete.
— Linear regression makes use of a specific set of assumptions about the data, known as the 'model', as follows: (a) there is an unknown true functional relationship $y = \alpha + \beta x$; (b) the x-values are fixed by the experimenter; (c) the y-values are experimental results and therefore subject to variation under repeated measurement; and (d) in simple regression a single unknown variance σ^2 describes the variation of the y-values around the true line.

Regression is sometimes loosely described as 'fitting the best straight line' to a set of points. There are in fact a number of ways of fitting a line to such a set of points, and which method is 'best' depends on both the intentions of the scientist and the nature of the data. Simple 'least-squares' regression is perhaps the method most widely used, and one that has especial relevance for calibration in analytical chemistry. Simple linear regression is based on a specific model of the data. It is worth taking the effort to understand how it works because if misapplied it can provide misleading results.

First, we assume that there is a true linear relationship between two variables x and y, namely $y = \alpha + \beta x$, which represents a straight line with slope β and intercept α. (The intercept is the value of y when x is zero.) Second, we assume that n values of the x-variable $x_1, x_2, x_3, \ldots, x_i, \ldots, x_n$ are fixed exactly by the experimenter. At each value x_i the corresponding value y_i is measured. As y_i is the result of a measurement, its value will not fall exactly on the line $y = \alpha + \beta x$, and will be different each time the measurement is repeated. (Remember that the x-values are exactly set: they are not the results of measurements.) Third, we assume that y_i is normally distributed with a mean value of $\alpha + \beta x_i$ and a variance of σ^2. This model can be written as

$$y_i = \alpha + \beta x_i + \varepsilon_i, \quad \varepsilon_i \sim N(0, \sigma^2).$$

The model is illustrated in Fig. 5.1.1. For the marked x-value, the corresponding y-value will be somewhere within the range of the indicated normal distribution. The point is more likely to fall closer to the centre of the distribution (on the line) than in the tails. Other x-values give rise to corresponding y-values, each one independently distributed around the line as $N(0, \sigma^2)$. When the scaffolding of the model is stripped away, we are left with the bare experimental points (Fig. 5.1.2). We have no information

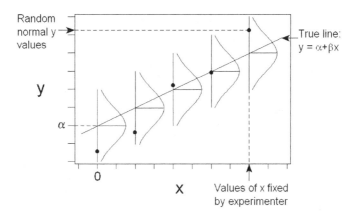

Fig. 5.1.1.   Model used for linear regression. The x-values are taken as fixed and the y-values subject to measurement variation.

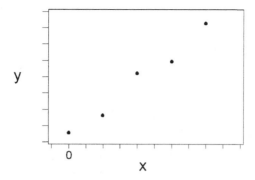

Fig. 5.1.2. A set of experimental points produced under the linear regression model (same points as in Fig. 5.1.1).

about the true relationship (even whether it is a straight line), nor the size of σ^2. The task of regression is to estimate the values of α and β as best we can, and check that these estimated values provide a plausible model of the data.

5.2 How Regression Works

Key points

— The regression line $y = a + bx$ is calculated from the points (x_i, y_i) by the method of least squares, that is, by finding the minimum value of the sum of the squared residuals.

— The regression coefficients are given by:

$$b = \sum_i (x_i - \bar{x})(y_i - \bar{y}) / \sum_i (x_i - \bar{x})^2, \quad a = \bar{y} - b\bar{x}.$$

— a and b are estimates of (but not identical with) the respective α and β.

— Regression is not symmetric: strictly speaking, we cannot exchange the roles of x and y in the equations.

Imagine that a straight line $y = a + bx$ is drawn through the data. The line could be regarded as a 'trial fit' (Fig. 5.2.1). We need to adjust the values of a and b in this equation until the line is a 'best fit' in some defined sense. The 'fitted points' $\hat{y}_i$ fall exactly on the line vertically above the points

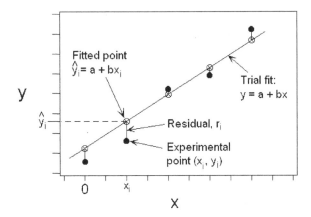

Fig. 5.2.1. Key aspects of regression, showing a trial fit with fitted values and the residuals. Same points as Fig. 5.1.1.

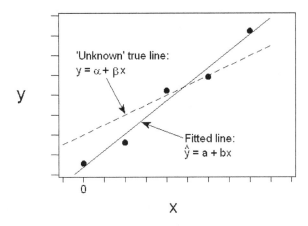

Fig. 5.2.2. Regression line (solid line) for the data shown in Figs 5.1.2 and 5.2.1. The 'true line' used to generate the data is shown dashed.

x_i, so that we have $\hat{y}_i = a + bx_i$. (In statistics, the notation $\hat{y}_i$ [spoken as 'y-hat'] implies that the quantity y_i is an estimate.) The residuals are defined as $r_i = y_i - \hat{y}_i$. We define the regression line by finding values of a and b that provide the smallest possible value of the sum of the squared residuals, Q. Thus we have

$$Q = \sum_i r_i^2 = \sum_i (y_i - \hat{y}_i)^2 = \sum_i (y_i - a - bx_i)^2.$$

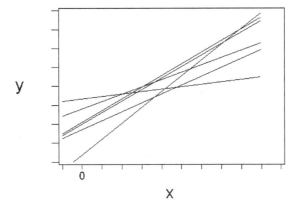

Fig. 5.2.3. Regression lines from repeated experiments under the conditions shown in Fig. 5.1.1, that is, with the same x-values, but a different random selection of y-values each time.

(Remember: in this equation the x_i and y_i are constants fixed by the experiment, and here we are treating a and b as variables.) We find the minimum value of Q by setting the first derivatives equal to zero, that is,

$$\frac{\partial Q}{\partial a} = 0, \quad \frac{\partial Q}{\partial b} = 0.$$

Solving these two simultaneous equations gives expressions for a and b, namely

$$b = \frac{\sum_i (x_i - \bar{x})(y_i - \bar{y})}{\sum_i (x_i - \bar{x})^2}, \quad a = \bar{y} - b\bar{x}.$$

Notes

- *The values of a and b are not the same as the respective unknown α and β (Fig. 5.2.2), but they are unbiased: if the experiment were repeated many times (that is, with new random y_i values each time, as in Fig. 5.2.3), the average values of a and b would tend towards α and β.*
- *The procedure is called the 'regression of y on x'. It is not symmetric: we get a different line if we exchange x and y in the equations for a and b. That would be an incorrect line, because the assumption that the x-values were error-free would be violated.*
- *x is called the 'independent variable' or the 'predictor variable'. y is called the 'dependent variable' or the 'response variable'.*

- *The procedure outlined above is called the 'method of least squares'. There are other procedures for fitting a line to data, but least squares is simple mathematically and meets most requirements so long as the data are produced in well-designed experiments.*
- *Always make the x, the error-free independent variable, the horizontal axis in x-y scatter plots.*

5.3   Calibration: Example 1

Key points
— Regression is suited to estimating calibration functions because we can usually regard the concentrations as fixed and the responses as variables.
— Unknown concentrations can be estimated from the transformed calibration function.

The process of calibrating an analytical instrument, by measuring the responses obtained from solutions of several known concentrations of the analyte, conforms closely to the assumptions of regression. The independent variable is the concentration of the analyte, accurately known because the

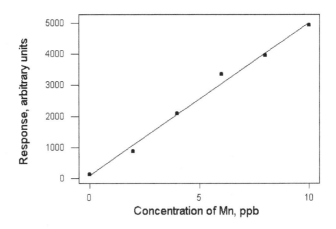

Fig. 5.3.1.   Calibration data (points) and regression line for manganese.

concentration is determined by gravimetric and volumetric operations only. It is a reasonable presumption that the concentrations of the analyte in the calibration solutions are known with a small relative uncertainty, often as good as 0.001. The dependent variable is the response of the instrument, and that is seldom known with a relative uncertainty less than 0.01.

Consider the following data obtained by constructing a short-range calibration for manganese determined by inductively-coupled plasma atomic emission spectrometry. Here we are making the responses the dependent variable y, and concentration the independent variable x. The regression line (Fig. 5.3.1) is given by the function $y = 75 + 496x$ and, visually, seems like a good fit to the data. We can use this function to estimate unknown concentrations in any other matrix-matched solution by inverting the equation to give $x = (y-75)/496$. A test solution providing a response of (say) 3000 units would indicate a concentration of manganese of $5.9\,\mu\mathrm{g\,l}^{-1}$.

We can also use the b value (the slope of the line) to convert response data for estimating the detection limit (according to the simple definition in §9.6). If we record $n > 10$ repeated responses when the concentration of the analyte is zero (that is, with a blank solution) and calculate the standard deviation s of these responses, the detection limit is given by $c_L = 3s/b$.

Other important features of the calibration function can be tested after regression. In practice we would need to check the validity of various assumptions underlying the regression and these items are considered in the following sections.

Concentration of Mn, ppb	Response, arbitrary units
0	114
2	870
4	2087
6	3353
8	3970
10	4950

Note

The dataset is available in the file named **Manganese1**.

5.4   The Use of Residuals

> **Key points**
> — Examining a plot of the residuals is an essential part of using regression.
> — If there is no lack of fit between the observations and the fitted model, the residuals should resemble a random sample of independent values from a normal distribution.

If the assumptions of regression are fulfilled and the measured data are truly derived from a straight line function, then we expect the residuals to behave very like a random selection from a normal distribution centred on zero. The variance of the residuals $s_{y\,|\,x}^2$ is given by

$$s_{y\,|\,x}^2 = \sum_i (y_i - \hat{y}_i)^2 / (n - 2).$$

Note that this expression is similar to the ordinary expression for variance, except that here the deviations are measured from the fitted values $\hat{y}_i$ (instead of from the mean $\bar{y}$) and there are now $n - 2$ degrees of freedom. (There are $n - 2$ degrees of freedom because we have calculated *two* statistics (a and b) from n pairs of observations.) This variance $s_{y\,|\,x}^2$ estimates σ^2 (see §5.1) if the regression line is a good fit to the data. The standard deviation of the residuals is of course the square root of the variance. So if we divide the residuals by $s_{y\,|\,x}$, these 'scaled residuals' should resemble a sample from standard normal distribution $N(0,1)$. This provides a useful method of checking visually whether the line produced by regression is an acceptable fit and whether the data plausibly conform to the assumptions underlying regression.

For the manganese calibration data we find that $s_{y\,|\,x} = 190$ and obtain Figs. 5.4.1 and 5.4.2 when (a) the residuals and (b) the scaled residuals are plotted against the x-values. The pattern of the residuals is seen to correspond to results that fall below or above the regression line. In other instances the deviations from the line may be too small to see on the plot of response against concentration, but they will always be apparent in the

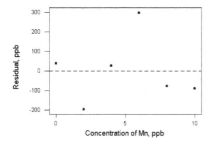

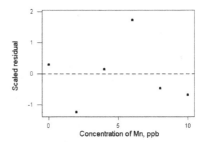

Fig. 5.4.1. Residuals from the linear regression of the manganese calibration data.

Fig. 5.4.2. Scaled residuals from the linear regression of the manganese calibration data.

residual plot. In this case, the successive residuals could plausibly be a random selection from a normal distribution, so we conclude that there is no reason to suspect any lack of fit. In other words, linear regression has provided a line that fits the data well. It is somewhat easier to see this from the scaled residuals, because all of the values fall within bounds of about ± 1.6. In a genuine $N(0, 1)$ distribution, we expect about 90% of data to fall within these limits on average.

Producing and considering residual plots is an essential part of using regression. It helps to avoid using inappropriate methods and making incorrect decisions. However, it is important not to over-interpret these plots when the number of data points is small, as in most examples from analytical calibration. Patterns indicating lack of fit or other problems (see §5.5) must deviate strikingly from a random, independent sample from a normal distribution before lack of fit is inferred from small numbers of residuals. Where possible, such inferences should be backed up by numerical tests of significance (see §5.10, 5.11).

Notes

- *The dataset is available in the file named **Manganese1**.*
- *The least squares calculation ensures that the mean of the residuals is exactly zero, so the residuals are not quite independent.*

5.5   Suspect Patterns in Residuals

Key points
— There are several ways in which a residual plot can deviate from that expected for a good fit.
— There are diagnostic patterns for outliers, lack of fit and non-uniform variance.

Where plots of residuals show patterns that deviate strongly from a random normal sequence, it is likely that the original data deviate from the assumptions of the regression model used, and that the results of the regression might be misleading. A different model might provide a more accurate and useful outcome. There are several patterns that analytical chemists should be aware of.

The first type of pattern (Figs. 5.5.1, 5.5.2) occurs when there is an outlying point among the data. This is apparent as a residual with a standardised value of 2.2. (This example is only marginally outlying.) The remaining residuals seem acceptable, although there is a negative mean. An outlier can bias the regression by drawing the regression line towards itself and thus may provide misleading information (§5.11). If they can be clearly identified as such, outliers should be deleted from calibration data. Statistical testing for outliers in calibration data is not simple, but applying Dixon's test (§7.3) or Grubbs' test (§7.4) to the residuals would probably provide a reasonable guide for testing a single suspect data point.

The second type of pattern occurs when the regression model does not fit the data properly. In the instance shown (Figs. 5.5.3, 5.5.4) a linear

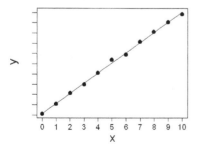

Fig. 5.5.1.   Data and regression line with a suspect point at $x = 5$.

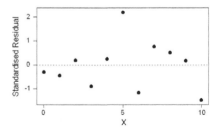

Fig. 5.5.2.   Standardised residuals from regression line in Fig. 5.5.1, showing a single suspect point at $x = 5$.

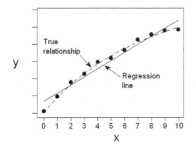

Fig. 5.5.3. Data and linear regression line showing lack of fit due to curvature in the true relationship. The lack of fit at very low concentrations could be seriously misleading.

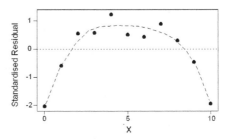

Fig. 5.5.4. Standardised residuals showing a strongly bowed pattern, indicating systematic lack of fit of the data to the regression line.

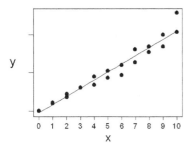

Fig. 5.5.5. Heteroscedastic data and regression line giving rise to residuals that tend to increase with x.

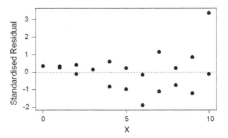

Fig. 5.5.6. Standardised residuals showing a heteroscedastic tendency: the residuals tend to increase with x. (They also suggest a possibly low bias in the regression line below about $x = 3$.)

regression has been applied to data with a curved trend, and the residuals show a corresponding bow-shaped pattern. It is important to avoid this type of lack of fit in calibration, because of the relatively large discrepancy at low concentrations between the regression line and the true trend of the data. The occurrence is dealt with by the use of a more complex regression model, such as a polynomial (§6.3, 6.4) or a non-linear model (§6.9, 6.10). The pattern may be unconvincing (but also prone to over-interpretation) if only a small number of points are represented. Statistical tests for non-linearity are therefore a useful adjunct to residual plots, and are discussed below (§5.10).

The third type of suspect residual pattern shows residuals that vary in size with the value of the independent variable. In the example illustrated (Figs. 5.5.5 and 5.5.6) there is a tendency for the residuals to increase in magnitude with increasing x. This phenomenon is called 'heteroscedasticity'. The tendency is contrary to the assumption of simple regression that the variance σ of the y-values is constant across the range of the independent variable x. The feature is common in analytical calibrations unless the range of concentration is quite small (for example, 1–2 orders of magnitude of concentration above the detection limit). Where several orders of magnitude of concentration are covered in a calibration set, heteroscedasticity may be pronounced. If simple regression (i.e., as discussed so far) is used on markedly heteroscedastic data, the resulting line will have a magnified uncertainty towards zero concentration (§6.7). This may have a disastrous effect on the apparent detection limit and the accuracy of results in this region (see Fig. 6.8.3). The correct approach to heteroscedastic data is to use a statistical technique called weighted regression (§6.7). As with other suspicious patterns in residuals, there is a tendency among the inexperienced to see random data as patterns when the number of points is small. The complication of using weighted regression is best avoided unless the need for it is completely clear.

5.6   Effect of Outliers and Leverage Points on Regression

Key points

— Outliers and leverage points can adversely affect the outcome of regression.

— They are often easy to deal with or avoid in analytical calibration.

There are two kinds of suspect data points that can adversely affect the outcome of regression, namely outliers and leverage points. Outliers are essentially anomalous in the value of the dependent variable, that is, in the y-direction in a graph (Fig. 5.6.1). They have the effect of drawing the fitted line towards the outlying point and thus rendering it unrepresentative of the other (valid) points. Because of this, regression should never be undertaken without a prior visual appraisal of the data or a retrospective residual plot. Extreme outliers will be immediately obvious and should be removed from the dataset. Marginal outliers are more difficult to deal with. Some statistical software packages give an indication of which points could reasonably be regarded as outliers. In any event, practitioners must avoid deleting the point with the largest residual without careful thought. That

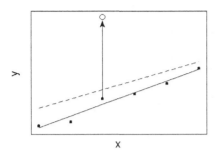

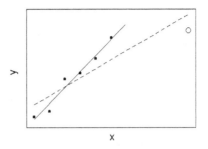

Fig. 5.6.1. Effect of an outlier on regression. When one of the original points (solid circles) is moved to an outlying position (open circle), the original regression line (solid) moves to an unrepresentative position (dashed line).

Fig. 5.6.2. Effect of a leverage point on regression. The well-spaced points (solid circles) give rise to the original regression line (solid). The extra leverage point (open circle) draws the regression line (dashed) to an unrepresentative position.

would always reduce the variance of the residuals, but would not improve the fit unless the point is a genuine outlier. In calibration it will normally be possible for the analyst quickly to check the data or the calibrators for mistakes, so it is probably better to repeat the whole calibration if an outlier is encountered.

Leverage points are anomalous in respect of the independent variable (Fig. 5.6.2). Because of their distance from the other points they can draw the fitted line towards them. Even if they are close to the same trend as the rest of the points, they will have an undue influence on the regression line. Again, leverage points should be treated with caution. They can easily be avoided in calibration. Again, some statistical software will identify points that have undue leverage.

5.7 Variances of the Regression Coefficients: Testing the Intercept and Slope for Significance

Key points
— The variances of the regression coefficients can be calculated from the data.
— They can be used to conduct significance tests on the coefficients.

Because the y-values in regression are variables (that is, subject to random error of measurement), the estimates a and b are also variables, and it is of interest to note how large that variability is. This gives us information about how much reliance we can place on the calculated values of a and b. We can obtain the required statistics from the data itself. Thus we have an expression for the variance of the slope b, namely:

$$\mathrm{var}(b) = s^2_{y\,|\,x}/\sum_i (x_i - \bar{x})^2,$$

and for the intercept a:

$$\mathrm{var}(a) = \mathrm{var}(b) \sum_i x_i^2/n.$$

The standard errors $se(b)$ and $se(a)$ are simply the respective square roots of these variances. Various null hypothesis about the coefficients can be tested for significance by using these standard errors.

We can test whether there is evidence to discount the idea that a calibration line plausibly passes through the origin (that is, that the instrumental response is zero when the concentration of the analyte is zero). For that purpose we consider the null hypothesis $H_0 : \alpha = 0$ by calculating the value of Student's t, namely $t = (a - \alpha)/se(a) = a/se(a)$ and the corresponding probability. A sufficiently low value (say $p < 0.05$) should convince us that we can safely reject the null hypothesis. (*Note*: a test for $H_0 : \alpha = 0$ is pointless unless there is a good fit between the data and the regression line: for example, Figs. 5.5.3 and 5.11.3 show a lack of fit situation where the intercept of the regression line differs obviously from the true trend of the data.)

Hypotheses about the slope b can likewise be tested by calculating $t = (b - \beta)/se(b)$ and the corresponding probability under various hypotheses about β. If we consider $H_0 : \beta = 0$, we are asking if there is any relationship at all between x and y, that is, whether the slope is zero (where the value of y is unaffected by the value of x). That circumstance is irrelevant in calibration, but might arise in exploratory studies of data, where we want to see which (if any) of a large number of possible predictors has an effect on the response (§6.5, 6.6). In calibration it is possible that we might want to compare an experimental value of b with a literature value b_{lit}, in which case we have $H_0 : \beta = b_{lit}$. Finally, we might consider $H_0 : \beta = 1$ in studies of bias over extended ranges, but in such instances we have to take care that the assumptions of regression are not seriously violated (see §5.12, 5.13).

For our manganese calibration data (§5.3) we find the following.

	Coefficient	Standard error	t	p
Intercept a	75.5	137.7	0.55	0.613
Slope b	496.37	22.74	21.83	0.000

The high p-value for the intercept means that there is no evidence to reject the idea that the true calibration line passes through the origin. A very low p-value for the slope is inevitable in analytical calibration and is of no inferential value.

Notes

- *The data used can be found in the file named* **Manganese1**.
- *Most computer packages give t-values and corresponding values of p alongside the estimates of the regression coefficients. The t-value usually relates to the null hypothesis that the respective coefficient has a zero population mean.*

5.8 Regression and ANOVA

> **Key point**
> — In regression, the variance of the y-values can be analysed into a component attributed to the regression and a residual component.

In regression the variation among the y-values can be split between the component due to the regression and that due to the residuals. This enables us to compare the relative magnitude of the components and make deductions about the success of the regression. The overall variance of the y-values is given by the normal expression for variance, namely:

$$\sum_i (y_i - \bar{y})^2 / (n - 1).$$

The variance of the residuals we have seen (§5.4) is

$$\sum_i (y_i - \hat{y}_i)^2 / (n - 2).$$

For the variance due to the regression (that is, of the fitted values $\hat{y}_i$ around $\bar{y}$), there is a denominator of one because there is only one degree of freedom remaining, namely $(n-1) - (n-2) = 1$, so the variance is simply $\sum_i(\hat{y}_i - \bar{y})^2$. Computer packages provide an ANOVA table composed as follows.

Source of variation	Degrees of freedom	Sum of squares	Mean square (variance)	F
Regression	1	$\sum_i(\hat{y}_i - \bar{y})^2$	$\sum_i(\hat{y}_i - \bar{y})^2$	
Residuals	$n-2$	$\sum_i(y_i - \hat{y}_i)^2$	$\sum_i(y_i - \hat{y}_i)^2/(n-2)$	$\dfrac{\text{Regression mean square}}{\text{Residual mean square}}$
Total	$n-1$	$\sum_i(y_i - \bar{y})^2$	$\sum_i(y_i - \bar{y}_i)^2/(n-1)$	

The value of F can be used as another test of a significant relationship between y and x, and is mathematically equivalent to testing $H_0 : \beta = 0$ with a t-test. Another statistic provided by ANOVA is designated R^2 and is the ratio of the regression sum of squares to the total sum of squares (usually expressed as a percentage). This can be thought of as the proportion of the variation in the y-values that is accounted for by the regression. It is numerically equal to the square of the correlation coefficient (§5.9) between the y-values and the fitted values. It can be used as a *rough* guide to the success of the regression, because a value approaching 100% means that most of the variation has been accounted for. But the statistic must be treated with caution: a model providing (say) a 99% value of R^2 is not necessarily better than one that provides a value of 95%, for the same reasons that apply to the correlation coefficient.

The example calibration data for manganese (§5.3) provides the following ANOVA table.

Source of variation	Degrees of freedom	Sum of squares	Mean square (variance)	F	p
Regression	1	17246922	17246922		
Residuals	4	144830	36207	476.3	0.0000
Total	5	17391751			

We see that a value of $R^2 = 99.2\%$. Such a high value is usual in analytical calibration unless a grossly inappropriate model has been used. Likewise, the high F-value and the corresponding small probability are typical of calibration and of no real interest.

5.9 Correlation

Key points

— Correlation is a measure of the relationship between two variables.
— A perfect linear relationship provides a correlation coefficient of exactly 1 or -1.
— The correlation coefficient is not a reliable indicator of linearity in calibration.

Correlation is a measure of relationship between two variables. It is related to, but distinct from, regression, but it is prone to misinterpretation unless great care is taken. Valid inferences that can be made with the help of the correlation coefficient are few in analytical science, and the statistic is best avoided. This section is essentially a warning. Unfortunately, some computer packages provide the correlation coefficient as routine as a by-product of regression, which give it a false appearance of applicability.

The correlation coefficient is defined as

$$ r = \frac{\sum_i ((x_i - \bar{x})(y_i - \bar{y}))}{\sqrt{\sum_i (x_i - \bar{x})^2 \sum_i (y_i - \bar{y})^2}}. $$

The value of r must fall between ± 1, regardless of the actual x- and y-values. Unlike regression, it is symmetric in x and y, the identical value of r being produced if the roles of x and y are interchanged in the equation.

When there is no relationship between the variables, that is, the y-values do not depend on the x-values in any way, the correlation coefficient will be zero. When the points fall exactly on a straight line the coefficient takes a value of $+1$ for lines with a positive slope or -1 for lines with a negative slope. For intermediate situations (points scattered at a greater or lesser

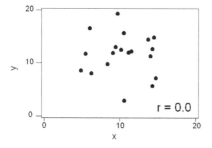

Fig. 5.9.1. Scatterplot of 20 points with zero correlation.

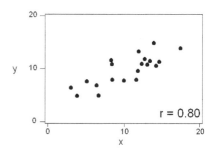

Fig. 5.9.2. Scatterplot of 20 points with a correlation coefficient of 0.80.

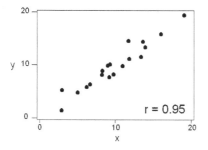

Fig. 5.9.3. Scatterplot of 20 points with a correlation coefficient of 0.95.

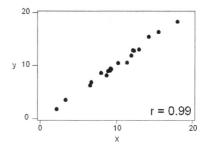

Fig. 5.9.4. Scatterplot of 20 points with a correlation coefficient of 0.99.

distance from a straight line) the coefficient takes values $0 < |r| < 1$. Some examples are shown in Figs. 5.9.1–5.9.4.

Some problems of interpreting r are as follows.

- Outliers have a strong effect on the value. In Fig. 5.9.5 we see the same data as in Fig. 5.9.2 but with an outlier added. The coefficient has increased from 0.80 to 0.92, despite the fact that the outlier does not lie on the same trend as the original data.
- Points with an exact functional relationship that is not linear do not necessarily give values of r above zero (Fig. 5.9.6).
- While points very close to a straight line provide a coefficient of almost 1 (or -1) by definition, *the converse is not true*. Points with $r \approx 1$ do not have to be scattered around a straight line. This kind of ambiguity is illustrated in Figs. 5.9.7 and 5.9.8, where points with a straight tendency and a curved tendency provide identical values of r. This ambiguity extends to values of r that are even closer to unity.

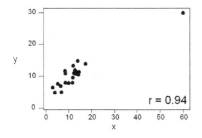

Fig. 5.9.5. Scatterplot of data with an outlier, which tends to increase the correlation coefficient.

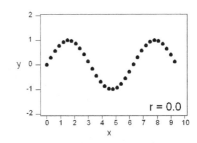

Fig. 5.9.6. Bivariate dataset with an exact functional relationship but zero correlation.

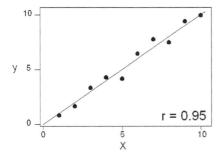

Fig. 5.9.7. Data with a linear trend and a correlation coefficient of 0.95.

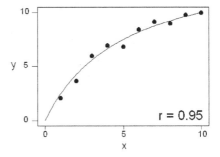

Fig. 5.9.8. Data with a non-linear trend and a correlation coefficient of 0.95.

All of this shows that r has little to commend it in the context of chemical measurement, and that a scatterplot is much more informative and nearly always to be preferred. It is particularly unfortunate when r is used as a test for sufficient linearity in calibration graphs in analytical science. It is quite possible to have a calibration with $r = 0.999$ and still demonstrate significant lack of fit to a linear function (§5.11). Moreover, it is dangerous to compare different values of r. For example, it would be wrong to say that a dataset with $r = 0.999$ was more linear than a set with $r = 0.99$. (Other tests for lack of fit *are* statistically sound: for example, the pure error test [§5.10] and polynomial fitting.)

Note

- *The use of r for the correlation coefficient must not be confused with the same symbol used for 'residual' or for repeatability.*

5.10   A Statistically-Sound Test for Lack of Fit

> **Key points**
> — Lack of fit between data and a regression line can be studied by making an independent estimate of the 'pure error'.
> — This requires replicated responses to be recorded.

Calibration lines that are straight or very close to straight occur often in analytical measurement, and such a calibration is highly beneficial in that it tends to reduce the uncertainty of predicted concentrations as well as being easiest to fit. Consequently, there is a tendency for analytical chemists to assume linearity and ignore small deviations from it. In many instances that is completely justified. However, fitting a straight line to data that have a curving trend will produce errors that are likely to be serious at the bottom end of the calibration. It is therefore of considerable importance to check whether calibration lines actually are straight, and if they are not, to determine the magnitude of the discrepancy. As a result, tests for linearity are written into procedures for method validation in many analytical sectors. Unfortunately, these tests are mostly based on the correlation coefficient and therefore are statistically unsound (see §5.9). Here we consider a test for lack of fit that is both sound and suitable for method validation. It requires an independent estimate of σ^2, which can be obtained by replicating some or all of the response measurements.

A general scheme for the independent estimation of σ^2 (called the 'pure error mean square' MS_{PE}) is given below. Suppose there are m different concentrations and the response at each concentration is measured n times. Then we have the following data layout from which we can calculate sums of squares and degrees of freedom.

Concentration	Repeated responses	Sum of squares	Degrees of freedom
x_1	$y_{11}, \ldots, y_{1i}, \ldots, y_{1n}$	$SS_1 = \sum_i (y_{1i} - \bar{y}_1)^2$	$n-1$
$\vdots$	$\vdots$	$\vdots$	$\vdots$
x_j	$y_{j1}, \ldots, y_{ji}, \ldots, y_{jn}$	$SS_j = \sum_i (y_{ji} - \bar{y}_j)^2$	$n-1$
$\vdots$	$\vdots$	$\vdots$	$\vdots$
x_m	$y_{m1}, \ldots, y_{mi}, \ldots, y_{mn}$	$SS_m = \sum_i (y_{mi} - \bar{y}_m)^2$	$n-1$
Totals	—	$SS_{PE} = \sum_j SS_j$	$m(n-1)$

Then we have:

$$MS_{PE} = SS_{PE}/(m(n-1)).$$

The sum of squares for lack of fit is the residual sum of squares SS_{RES} minus the pure error sum of squares, so the mean squares due to lack of fit is

$$MS_{LOF} = (SS_{RES} - SS_{PE})/(n-2).$$

The test statistic is

$$F = MS_{LOF}/MS_{PE},$$

which has $(n-2)$ and $m(n-1)$ degrees of freedom.

This looks a bit formidable, but the calculations are quite straightforward and will always be carried out by computer.

5.11   Example Data/Calibration for Manganese

> **Key points**
> — Residual plots and tests for lack of fit serve to detect non-linearity in a calibration plot.
> — These are more reliable tests than a consideration of the correlation coefficient.
> — Outlying results can perturb the conclusions for such tests.

The calibration data for manganese previously considered are actually part of a larger set of duplicated results at each concentration. The data set up to concentration 10 ppb is given here.

Concentration, ppb	Response 1	Response 2
0	114	14
2	870	1141
4	2087	2212
6	3353	2633
8	3970	4299
10	4950	5207

The regression and residual plots are shown in Figs. 5.11.1 and 5.11.2. By visual judgement there is no suggestion of lack of fit in the residual plot.

The following simple calculations lead to the pure error estimate. $d_a^2 + d_b^2$ is the pure error sum of squares for duplicated results. (This data layout is for demonstration purposes only: the calculations will always be carried out by computer.)

x	y_a	y_b	$\bar{y}$	$d_a = y_a - \bar{y}$	$d_b = y_b - \bar{y}$	$d_a^2 + d_b^2$	DOF
0	114	14	64.0	50.0	−50.0	5000	1
2	870	1141	1005.5	−135.5	135.5	36721	1
4	2087	2212	2149.5	−62.5	62.5	7813	1
6	3353	2633	2993.0	360.0	−360.0	259200	1
8	3970	4299	4134.5	−164.5	164.5	54121	1
10	4950	5207	5078.5	−128.5	128.5	33025	1
Totals	—	—	—	—	—	395878	6

This gives rise to the following ANOVA table. The probability associated with the lack of fit F statistic is high and confirms the absence of significant lack of fit.

Source of variation	Degrees of freedom	Sum of squares	Mean square (variance)	F	p
Regression	1	35608623	35608623	819.34	0.000
Residual error	10	434603	43460		
Lack of fit	4	38725	9681	0.15	0.958
Pure error	6	395878	65980		
Total	11	36043226			

If we now consider the full dataset, up to a concentration of 20 ppb we have the following data.

Concentration ppb	Response 1	Response 2	Concentration ppb	Response 1	Response 2
0	114	14	12	5713	5898
2	870	1141	14	6496	6736
4	2087	2212	16	7550	7430
6	3353	2633	18	8241	8120
8	3970	4299	20	8862	8909
10	4950	5207	—	—	—

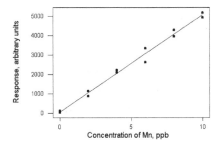

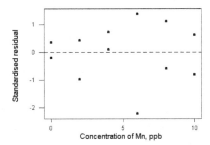

Fig. 5.11.1. Calibration data for manganese, with duplicated results for response.

Fig. 5.11.2. Residuals from the simple calibration of the manganese data. There is no suggestive pattern (although there is a single suspect point).

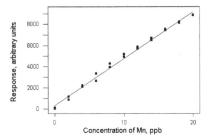

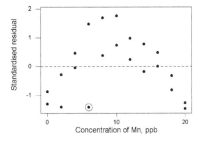

Fig. 5.11.3. Calibration data for manganese, up to 20 ppb.

Fig. 5.11.4. Residuals (points) from the calibration using simple regression. There is a strong visual suggestion of lack of fit, which is perhaps weakened by an outlying point (encircled).

The calibration plot (Fig. 5.11.3) and residual plot (Fig. 5.11.4) suggest a slight curvature in the relationship, but it is difficult to be sure because of the relatively large variability in the responses at each level of concentration.

On completion of the linear regression with the test for lack of fit, we have the following ANOVA table. The probability of 0.096 associated with the lack of fit test is low enough to substantiate the visual appraisal, although not significant at the 95% level of confidence. The probability may have been affected unduly by one apparently discrepant value of response at 6 ppb concentration, which inflates the estimate of pure error variance and thereby reduces the F-value. If this discrepant value is deleted before the test, the lack of fit becomes significant with $p = 0.007$.

Source of variation	Degrees of freedom	Sum of squares	Mean square (variance)	F	p
Regression	1	173714191	173714191	2631.62	0.000
Residual error	20	1320205	66010		
Lack of fit	9	862790	95866	2.31	0.096
Pure error	11	457416	41583		
Total	21	175034397			

Notes

- The datasets used in this section can be found in files named **Manganese2** and **Manganese3**.

- The pure error test for lack of fit takes no account of the order in which the residuals are arranged. The test statistic would have the same result if the order of the residuals were randomised. This has two corollaries: (a) non-linearity may be somewhat more likely than suggested by the p-value; and (b) lack of fit, if detected, may have a cause other than non-linearity. Hence it is essential to examine a residual plot.

- The coefficient of correlation between the concentrations and the responses in dataset **Manganese3** is $r = 0.996$. This would often incorrectly be taken as a demonstration of linearity in the calibration function. This example substantiates the previous comments (§5.9) about the shortcomings of the correlation coefficient as a test for linearity.

5.12   A Regression Approach to Bias Between Methods

Key points

— Linear regression can be used to test for bias between two analytical methods by using them in tandem to analyse a set of test materials.

— The inference will be safe so long as the results of the more precise method are used as the independent (x) variable.

The results of two analytical methods can be compared by using both of them to analyse the same set of test materials. When the range of concentrations determined is small, a t-test on the differences between paired results is usually appropriate (§3.8, 3.9). When the range is greater, a number of possible outcomes are possible, and the data can sometimes be analysed by regression or a related method. The regression approach is safe so long as the variance of the independent (x) variable is somewhat smaller than that of the dependent (y) variable. If the variances are comparable, or that of y exceeds that of x, regression is likely to provide misleading statistics, because a basic assumption of regression (invariant x-values) has been notably violated. Such instances can, however, be readily managed by

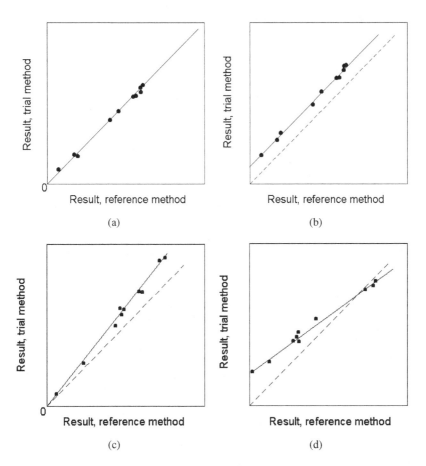

Fig. 5.12.1. Possible outcomes of experiments for the comparison of two methods of analysis, showing the regression line (solid) and the theoretical line of no bias (dashed). Each point represents the two results on a particular material. (a) No bias between the methods. (b) Translational bias only between the methods. (c) Rotational bias only between the methods. (d) Both rotational and translational bias between the methods.

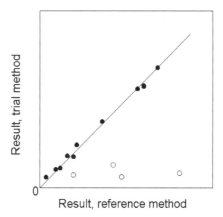

Fig. 5.12.2. A complex outcome of an experimental comparison between analytical methods where the test materials fall into two subsets, one subset with no bias (solid circles) and the remainder with a serious (but not statistically characterised) bias (open circles).

the use of a more complex method known as functional relationship fitting, which is beyond the scope of the present book.

Bias between two analytical methods can adopt one of several different 'styles', the most common of which are well modelled by linear regression in paired experiments. If both methods gave identical results (apart from random measurement variation, of course) the trend of results would follow a line with zero intercept and a slope of unity, that is, $y = \alpha + \beta x$ where $\alpha = 0$, $\beta = 1$ (Fig. 5.12.1a). As the methods are intended to address the same measurand, the two methods should produce results quite close to the ideal model ($y = x$). An obvious statistical approach is to test the hypotheses $H_0 : \alpha = 0$ vs $H_A : \alpha \neq 0$ and $H_0 : \beta = 1$ vs $H_A : \beta \neq 1$.

Figure 5.12.1(b) shows a dataset where $\alpha \neq 0$, $\beta = 1$. This type of bias is called 'translational' or 'constant' bias, and is commonly associated with baseline interference in analytical signals. Another common type of bias is characterised by $\alpha = 0$, $\beta \neq 1$, that is, the slope has departed from unity. This style of bias (Fig. 5.12.1c) is called 'rotational bias' or proportional bias. It is quite possible for both types of bias to be present simultaneously, that is, $\alpha \neq 0$, $\beta \neq 1$ (Fig. 5.12.1d).

In some instances a more complex behaviour may be seen. Fig. 5.12.2 shows an example where the results from the majority of the test materials follow a simple trend line, but a subset of test materials give results that show quite different behaviour.

Notes and further reading

- *Ripley, B. and Thompson, M. (1987). Regression Techniques for the Detection of Analytical Bias. Analyst, **112**, pp. 377–383.*
- *'Fitting a linear functional relationship to data with error on both variables'.(March 2002). AMC Technical Briefs No. 10. Free download via www.rsc.org/amc.*
- *Software also available for Excel and Minitab. Free download via www.rsc.org/amc.*

5.13 Comparison of Analytical Methods: Example

Key point

— Regression shows the bias between a rapid field method for the determination of uranium and an accurate laboratory method.

As an example we consider the determination of uranium in a number of stream waters by a well-established laboratory method and by a newly-developed rapid field method. The purpose is to test the accuracy of the field method, regarding the laboratory method as a reference point for accuracy. The results, in units of $\mu g\,l^{-1}$ (ppb) are as follows

Site code	Field result	Laboratory result
1	24	19
2	8	8
3	2	2
4	1	1
5	10	9
6	26	23
7	31	27
8	17	11
9	4	4
10	0	0
11	6	4
12	40	26
13	2	3
14	2	4

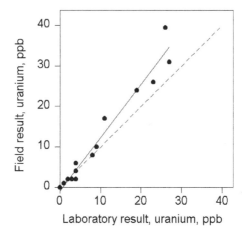

Fig. 5.13.1. Comparison between two methods for the determination of uranium in natural waters, showing the fitted regression line (solid) and the theoretical line for zero bias between the methods (dashed line). Each point is a separate test material.

Figure 5.13.1 shows the field results plotted against the laboratory results. We see that the trend of the results deviates from the theoretical line for unbiased methods where both methods give the same result (apart from random measurement variation). It seems reasonable to estimate this trend by linear regression of the field results against the laboratory results. This action can be justified in this instance because the precision of the laboratory method is known to be small in comparison with that of the field method.

The results of the regression are as follows. We see that the intercept is not significantly different from zero ($p = 0.412$), but the slope of 1.32 is clearly significantly different from unity, which is the slope required for no bias of any kind between the methods. The field method is giving results that are on average 1.32 times greater than the established laboratory method.

Predictor	Coefficient	Standard error	H_0, H_A	t	p
Intercept (a)	-0.944	1.111	$\alpha = 0, \alpha \neq 0$	0.85	0.412
Slope (b)	1.32065	0.08120	$\beta = 1, \beta \neq 1$	3.95	0.001
$s_{y \mid x} = 2.82$	$R^2 = 95.7\%$				

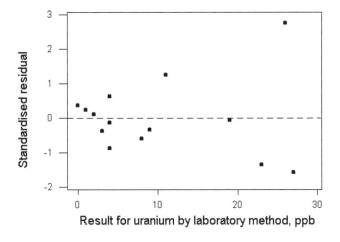

Fig. 5.13.2.   Residuals from the regression.

Notes

- *The dataset used in this section is in a file named **Uranium**.*
- *It is interesting to observe that the residuals of the regression (Fig. 5.13.2) tend to suggest a small degree of heteroscedastic variation in the results of the field method. Weighted regression should ideally have been used for this exercise (§6.7), although in this instance it would not have changed the interpretation of the data. In any event the estimation of the weights would have called for assumptions outside the data.*

Chapter 6

Regression — More Complex Aspects

This chapter examines some more complex aspects of regression in relation to analytical applications, specifically polynomial and multiple regression, weighted regression and non-linear regression.

6.1 Evaluation Limits — How Precise is an Estimated x-value?

Key points

— 'Evaluation limits' around concentrations estimated from calibration graphs can be calculated from the data and are often unexpectedly wide.
— Evaluation limits give rise to an alternative way of thinking about detection limits.

Values of concentration x' in unknown solutions, estimated from calibration lines, are subject to two sources of variation, first, the variation in the position of the regression line and second, the variation in the new measured response y'. The way that these two variances interact is shown schematically in Fig. 6.1.1. (Remember that y' and x' are not part of the calibration data.) The resultant variation in x' is, at first encounter, surprisingly large, but the reason for this is apparent from the diagram.

Confidence intervals around $x' = (y' - a)/b$ are given by $x' \pm t s_{x'}$ with an appropriate value of t, where

$$s_{x'} = \frac{s_{y|x}}{b} \sqrt{\frac{1}{m} + \frac{1}{n} + \frac{(y' - \bar{y})^2}{b^2 \sum_i (x_i - \bar{x})^2}}, \qquad (6.1.1)$$

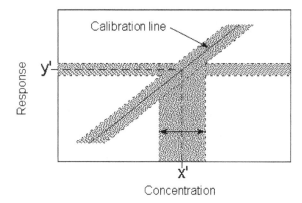

Fig. 6.1.1. Estimation of an unknown concentration x' from a response y' and an established calibration function. The shaded areas illustrate schematically the uncertainties in y', the calibration function (solid line) and x'.

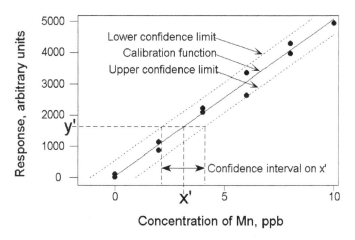

Fig. 6.1.2. Evaluation interval calculated from the manganese data (points), showing the 95% confidence interval around an estimated value of concentration.

and where y' is the mean of m observations, and the regression line is based on n pairs of $x - y$ data. The confidence limits around x' can be calculated for any value of y', and shown as two continuous lines which are gentle curves (see Fig. 6.1.2), which shows the curves calculated from the manganese calibration data. This procedure looks complex, but all of the statistics stem from the calibration data, and the limit lines can be calculated rapidly by computer. The manganese data shows that an unknown

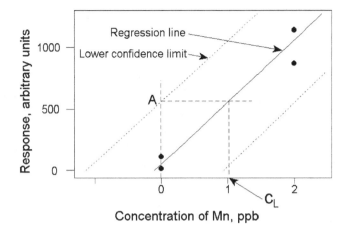

Fig. 6.1.3. Part of previous graph around the origin, illustrating how a detection limit can be estimated. At point A, the lower confidence limit intersects the zero concentration line. Concentrations below the corresponding detection limit c_L are not significantly greater than zero.

solution producing a response of (say) 1600 units provides an estimated concentration of manganese of 3.1 ppb with 95% confidence limits of 2.1 and 4.1.

Assuming that an appropriate model has been fitted, this provides a useful way of estimating of the contribution that stems from the calibration and evaluation procedures to the total uncertainty of the measurement result. It also provides us with an alternative means of conceptualising detection limits (Fig. 6.1.3). If we think about the lower confidence limit for a continuously diminishing concentration x', at some response (A in Fig. 6.1.3) the limit intersects the zero concentration line and, at that point and below, the estimated concentration is not significantly greater than zero. In this region we are not sure that there is *any* of the analyte present in the test solution.

Notes

- *The dataset used in the example diagrams are in the file named* **Manganese2**.
- *Terminology in this area is not stabilised. Here we call the limits 'evaluation limits', others call them 'inverse confidence limits' or 'fiducial limits'.*

- *Many textbooks correctly state that Eq. (6.1.1) is approximate. However, the error involved in the approximation is negligible in any realistic analytical calibration.*

6.2 Reducing the Confidence Interval Around an Estimated Value of Concentration

---

Key points
— Reducing the uncertainty in an estimated concentration is addressed by concentrating on the largest term in Eq. (6.1.1).
— The largest term is often $1/m$, where m is the number of replicate measurements of response for the unknown solution.

If the uncertainty in an estimated x' is too large, the equation for $s_{x'}$ above allows us to see which aspect of the calibration and evaluation needs attention, by comparing the separate magnitudes of the three terms under the square root sign. The magnitude of $1/m$ can be reduced by increasing m, the number of results averaged to obtain y'. This is normally the most effective strategy. Likewise, the magnitude of $1/n$ can be reduced by increasing n, the number of calibrators, and this also has the effect of reducing the value of t, which depends on the degrees of freedom, $(n-2)$.

The third term $\dfrac{(y' - \bar{y})^2}{b^2 \sum_i (x_i - \bar{x})^2}$ disappears when $y' = \bar{y}$ (usually around the centre of the calibration line) and, in most instances of calibration, is small compared with the other terms. (This is why the confidence limits usually look like straight lines parallel to the calibration function: strictly they are gentle curves.) It could be reduced somewhat by moving the calibrators to the extreme of the existing concentration range or by increasing the range of the existing calibration.

Common sense is necessary here: there is no point in worrying about the magnitude of errors contributed by calibration/evaluation if errors introduced at other levels of the analytical procedure exceed them to any extent. This is usually the case in chemical analysis, except at concentrations near the detection limit, where calibration errors assume a dominant magnitude.

6.3 Polynomial Regression

Key points
— Polynomial regression allows us to construct models that can fit data with a curved trend.
— It is sometimes suitable for calibration data, with models up to order two (i.e., with squared terms).
— Models of order higher than two are nearly always unsuitable for analytical calibrations, and give unreliable extrapolations.

When calibration data $x_i, y_i, (i = 1, \ldots, n)$ (or any other data) fall on a curved trend it is often possible to account for them successfully with a polynomial model which takes the form

$$y_i = \beta_0 + \beta_1 x_i + \beta_2 x_i^2 + \beta_3 x_i^3 + \cdots + \varepsilon_i, \quad \varepsilon \sim N(0, \sigma^2).$$

Notice here that the coefficients β are distinguished by a subscript that corresponds with the power to which the predictor variable is raised. The intercept is now called β_0 instead of α, so we can use a compact notation for the model, $y_i = \sum_{j=0}^{m} \beta_j x^j + \varepsilon_i$, where m is order of the polynomial, the highest power used. (Note that $\beta_0 x^0 = \beta_0$.) The least-squares coefficients can be calculated by an extension of the procedure used in §5.2 so long as $n > m + 2$, but in practice it is wise to use $n > 2m$.

It is very unlikely that a power of greater than three would be suitable for analytical calibration purposes. We can see this clearly in Fig. 6.3.1,

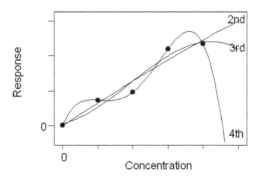

Fig. 6.3.1. Various polynomial fits to five calibration data points.

which shows fits, of up to order four, to five data points. The straight line fit is omitted for clarity. The second order 'quadratic' curve (part of a parabola) is a better fit to the points (in a least-squares sense) and provides a plausible calibration line. We find this plausible because we often observe slightly curved calibrations in practice and can usually account for them in terms of known physical processes taking place in the analytical system. The third order curve (part of a cubic equation) is still closer to the points, but now the fit is less plausible. We would be unhappy to use a calibration curve like that because there is an inflection in the curve that would be difficult to account for by physical theory. Finally, the fourth order fit passes exactly through each point, but is obviously nonsensical as a calibration.

In general, as the order of the fit is increased, the variance of the residuals becomes smaller. However, there is a point beyond which the improvement in fit is meaningless. There are statistical tests that can identify that point, but common sense is usually good enough in analytical calibration. A suitable procedure for analytical calibration is the following.

1. Determine at least six equally spaced calibration points.
2. Try a first order (linear) fit.
3. Examine the resulting residual plot.
4. If there is a lack of fit through curvature, try a quadratic (order two) fit.
5. Examine the new residual plot.
6. If there is still lack of fit, abandon the attempt to fit a polynomial.

All of these operations are easily accomplished in statistical packages.

While polynomials can be used satisfactorily to model calibrations with slight curves and short ranges, the fact remains that they are often inherently the wrong shape to describe the physical processes going on in an analytical system. This incompatibility is likely to become apparent in even small extrapolations. Consider the calibration data shown in Figs. 6.3.2–6.3.4. The graphs show various order fitted lines extrapolated, with the corresponding 95% confidence intervals. The first order fit shows the uncertainty in the extrapolation remains reasonably small. The higher

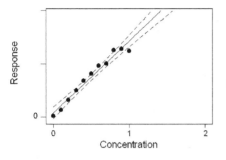

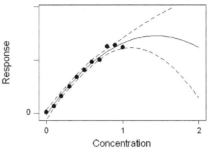

Fig. 6.3.2. Linear fit to calibration data, with 95% confidence interval.

Fig. 6.3.3. Second order fit to calibration data, with 95% confidence interval.

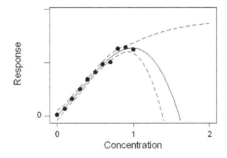

Fig. 6.3.4. Third order fit to calibration data, with 95% confidence interval.

order fits model that data rather more closely, but show extrapolations with a strongly curved trend and immediately very wide confidence interval. Extrapolation is very risky because we cannot infer the correct shape of the relationship from the data but only look for lack of fit within the range.

Note

- *In technical mathematical terminology, these polynomial models are all called 'linear', even though they describe curves: they are linear in the coefficients. There is another class of models technically called 'non-linear' that are more difficult to fit and discussed briefly in §6.9 and §6.10.*

6.4   Polynomial Calibration — Example

Key point

— Polynomial regression order two (quadratic) provides a good fit for the manganese data.

If we re-examine the complete manganese data from §5.11 by using quadratic regression, we obtain the following output, residual plot and analysis of variance table. The regression equation is:

$$\text{Response} = 3.03 + 550.24 \times c_{Mn} - 5.297 \times c_{Mn}^2,$$

where c_{Mn} is the concentration of manganese. The table of coefficients is as follows.

Predictor	Coefficient	Standard error	t	p
Intercept	3.03	91.64	0.03	0.974
Mn	550.24	21.32	25.81	0.000
Mn squared	−5.297	1.027	5.16	0.000
	$s_{y\mid x} = 170.1$	$R^2 = 99.7\%$		

We see that the squared term is highly significant in the test for H_0 : $\beta_2 = 0$, so we are disposed to think that the quadratic model will have improved the fit.

Source of variation	Degrees of freedom	Sum of squares	Mean square (variance)	F	p
Regression	2	174484616	87242308	3015.03	0.000
Residual error	19	549780	28936		
Lack of fit	8	92365	11546	0.28	0.960
Pure error	11	457416	41583		
Total	21	175034397			

The residual plot (Fig. 6.4.1) shows no trace of lack of fit (although the suspect value is now more obviously an outlier), and there is no apparent lack of fit near the origin. The analysis of variance shows no significant lack

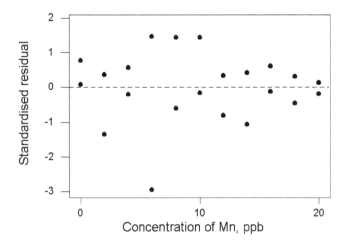

Fig. 6.4.1. Standardised residuals after carrying out quadratic regression on the manganese calibration data.

of fit, with $p = 0.96$. Quadratic regression therefore provides a good model for this particular calibration.

Note

- *The dataset used in this section is named **Manganese3**.*

6.5 Multiple Regression

Key points

— Multiple regression allows the exploration of datasets with more than one predictor variable.

— The technique has applications in analytical calibration in multivariate methods such as principal components regression.

— It is also useful in general exploratory data analysis.

Multiple regression is used when two or more independent predictors combine to determine the size of a response variable. The dataset layout is as follows, where the first subscript on each value of x indicates a separate variable.

Response variable	Predictor variables			
	1	2	...	m
y_1	x_{11}	x_{21}	...	x_{m1}
y_2	x_{12}	x_{22}	...	x_{m2}
y_3	x_{13}	x_{23}	...	x_{m3}
$\vdots$	$\vdots$	$\vdots$	...	$\vdots$
y_n	x_{1n}	x_{2n}	...	x_{mn}

The statistical model is therefore

$$y_i = \beta_0 + \beta_1 x_{1i} + \beta_2 x_{2i} + \beta_3 x_{3i} + \cdots + \varepsilon_i, \quad \varepsilon \sim N(0, \sigma^2),$$

and the equations to calculate the least squares estimates b_j of the coefficients β_j are derived in a manner similar to those for simple regression. As with polynomial regression, we should aim to have $n > 2m$ at least to obtain stable results.

Regression with two independent variables, that is, $\hat{y} = b_0 + b_1 x_1 + b_2 x_2$ can be readily visualised by means of perspective projections, and we can see such a model represented by a tilted plane OABC in the three-dimensional representations (Figs. 6.5.1, 6.5.2). When $x_2 = 0$ we have the simple regression $\hat{y} = b_0 + b_1 x_1$ shown as line OA. Likewise, when $x_1 = 0$ we have $\hat{y} = b_0 + b_2 x_2$ shown as line OC. At non-zero values of both x_1, x_2, values of $\hat{y}$ are represented by points on the plane OABC. (Note: in these two figures the value of b_0 is zero, but generally this will not be so.) When regression is executed on a set of points such as shown in Fig. 6.5.3, the residuals are the vertical distances (that is, in the y-direction) from the points to the fitted plane (Fig. 6.5.4). As in simple regression, it is essential

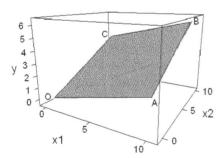

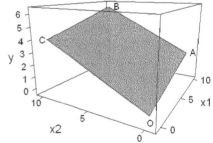

Fig. 6.5.1. Representation of a fitted function of two predictor variables as a tilted plane OABC in a three-dimensional space.

Fig. 6.5.2. Representation of a fitted function of two predictor variables as a tilted plane OABC in a three-dimensional space. (Same as Fig. 6.5.1, but with the origin at the bottom right corner.)

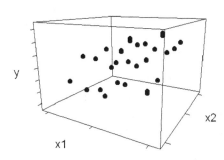

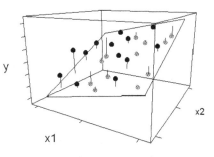

Fig. 6.5.3. Swarm of data points in three-dimensional space.

Fig. 6.5.4. Data points as in Fig. 6.5.3 and a fitted regression model (plane), showing the residuals. Points above the plane are shown in black, below the plane in grey.

good practice to examine the residuals, by plotting them against the value of each predictor variable in turn.

Multiple regression plays an essential part in the more complex types of calibration (for example in principal components regression [PCR]), but is perhaps more often used in exploratory data analysis. In the latter instance, it is important to check that the predictor variables used are not strongly correlated. (In PCR the predictors have zero correlation by definition.) If such correlation exists, the regression model may be unstable and the coefficients misleading. In other words, we might get very different results if two randomly selected subsets of the data were both separately treated by multiple regression.

6.6 Multiple Regression — An Environmental Example

Key points

— Multiple regression has been used to explore possible relationships between the concentration of lead in garden soils and predictors: (a) distance from a smelter; (b) the age of the house; and (c) the underlying geology.

— A preliminary regression gave a promising outcome but the residuals showed a lack of fit to distance.

— Replacing the distance predictor by a negative exponential transformation gave an improved fit with no significant lack of fit.

The data listed below show lead concentrations found in the soil of 18 gardens from houses in the vicinity of a smelter. Also listed for each garden are some corresponding environmental factors that might serve to explain the lead concentrations. The explanatory factors are as follows.

- The distance of each house from the smelter.
- The age of the house in years.
- The geological formation underlying the garden. (This is simply identified by codes 1 or 2 which show which one of two rocks is present.)

The task is to try multiple linear regression with lead concentration as the dependent variable and distance, age and type of geology as the predictors (independent variables). Note that while the geology code can only take one of two values, we can still use this variable as a predictor in multiple regression.

Lead, ppm	Distance, m	Age, Yr	Geology	Lead, ppm	Distance, m	Age, Yr	Geology
10	9609	22	1	136	6064	166	2
25	9283	58	1	146	3895	25	1
69	7369	61	2	164	4051	74	1
79	9887	132	2	170	6827	187	2
86	9887	118	1	184	4466	149	2
94	6130	121	1	198	4919	199	1
100	8328	125	1	201	6821	295	2
132	4790	42	2	219	3598	77	2
132	6248	139	2	275	3438	84	2

The first stage is to ensure that the intended predictors are not strongly correlated. The correlation matrix is as follows.

	Distance	Age
Age	0.040	
Geology	−0.243	0.298

All of these coefficients are small and not significant, so the three variables can be safely used in the intended regression. The next stage is to scan scatter plots (Figs. 6.6.1–6.6.3) of lead concentration against each variable in turn to see if the data are consistent with the proposed model. We can clearly see that the concentration of lead decreases with increasing distance, as would be expected (Fig. 6.6.1). There are no obvious trends suggesting non-linearity and no suspect points that might have an undue influence on the regression. There is no clear relationship between lead concentration

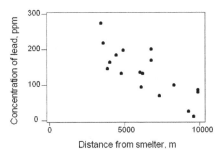

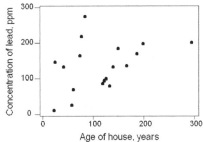

Fig. 6.6.1. Lead concentrations plotted against distance from the smelter ($r = -0.80$; $p < 0.0005$).

Fig. 6.6.2. Lead concentrations plotted against the age of the house ($r = 0.39$; $p = 0.11$).

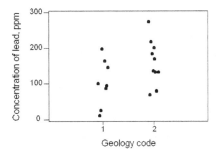

Fig. 6.6.3. Lead concentrations plotted against code for the underlying geology.

and the age of the house (Fig. 6.6.2), and the correlation is small and not significant. However, we must not be tempted to omit the age of the house from the regression at this stage. Quite often, real relationships are obscured by the influence of other variables. Figure 6.6.3 illustrates the effect of geology. In this diagram it seems that the mean of the results for the gardens coded as 2 is higher than the results for gardens coded as 1, but it is not clear whether the difference is significant. (Notice that the data points have been 'jittered' slightly in the x-direction in Fig. 6.6.3 so that overlapping points in the y-direction are separated.) There is no reason not to conduct the multiple regression with all three variables.

The outcome of the multiple regression is shown in the table. The R^2 value (proportion of variance accounted for by the regression) is 83%, which is very good for an environmental study. Both the distance from the smelter and the age of the house give values of t that are significantly different from zero at the 95% level of confidence. It is noteworthy that the age of the house is significant in the multiple regression when it seemed unpromising as a predictor when considered alone. The underlying geology apparently has no significant effect on the lead content of the soil, as the p-value is high.

Predictor	β_j	se(β_j)	$t = \beta_j/\text{se}(\beta_j)$	p
Constant	221.23	37.00	5.98	0.000
Distance	−0.024179	0.003494	−6.92	0.000
Age	0.3835	0.1146	3.35	0.005
Geology	15.63	16.00	0.98	0.345

While this outcome is promising, we now have to examine the residuals for lack of fit or other features that might throw doubt on the suitability of the first regression. The residuals plotted against predictors are shown in Figs. 6.6.4 and 6.6.5.

The residuals plotted against the age of the house display no obvious deviation from a random sample except for one suspect value (roughly at 90 years). Plotted against the distance from the smelter, however, there is a distinct suggestion of a curved trend in the residuals, showing that the linear representation was not completely adequate. Intuition supports this outcome, because we would expect the lead contamination to fall with

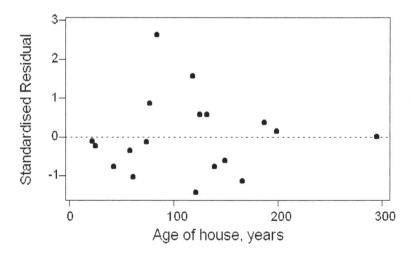

Fig. 6.6.4.   Residuals plotted against the age of the house.

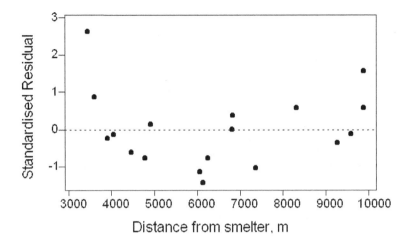

Fig. 6.6.5. Residuals plotted against the distance from the smelter.

distance from the smelter but in a roughly exponential manner. Notice that there is no exact way of using the pure error as a test for lack of fit, because there are no replicated results.

A possible improvement on the modelling might be obtained by transforming the distance variable d in some way that provides a suitably curved function, and using that as a predictor in a new multiple regression. We can try an exponential decay to give the new variable $\exp(-d/1000)$. (The division by 1000 is simply to provide results of a handy magnitude.) We now repeat the multiple regression and obtain the following results.

Predictor	β_j	$\mathrm{se}(\beta_j)$	$t = \beta_j/\mathrm{se}(\beta_j)$	p
Intercept	−7.31	14.03	−0.52	0.611
$\exp(-d/1000)$	6096.1	446.0	13.67	0.000
Age	0.63813	0.06815	9.36	0.000
Geology	14.661	8.848	1.66	0.120

The variance accounted for (R^2) is now 95% which is exceptionally high for an environmental study. Both the transformed distance and the age of the house are still highly significant predictors: the geology as a predictor remains not significant at the 95% level of confidence, but the p-value of 0.12 shows that its influence is not implausible, and a relationship

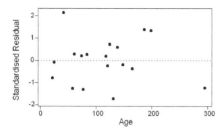 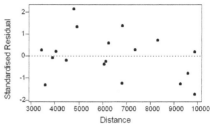

Fig. 6.6.6. Residuals from the second regression (with transformed distance) plotted against age of the house.

Fig. 6.6.7. Residuals from the second regression (with transformed distance) plotted against distance from the smelter.

might be revealed by a larger or more focused study. None of the residual plots from the new regression (Figs. 6.6.6 and 6.6.7) now show any obvious deviation from a random pattern. (Notice that the lead residuals have been plotted against the original distance measure rather than the transformed value, but that merely spaces them conveniently: it is not essential.)

Note

- *The dataset for this example is found in the file named **Lead**.*

6.7 Weighted Regression

> **Key points**
> — Weighted regression is designed for heteroscedastic data, i.e., the variance of the *y*-value varies with the *x*-value, as opposed to the assumption of uniform variance for simple regression.
> — Using simple regression where weighted regression should be used can have a deleterious effect on the statistics.
> — Analytical calibrations are often heteroscedastic, and weighted regression is often beneficial.
> — Weighted regression is a standard part of statistical packages.

Quite often in analytical calibration the range of the concentration of the analyte extends over several orders of magnitude. In such instances we usually find that the variance of the analytical response increases

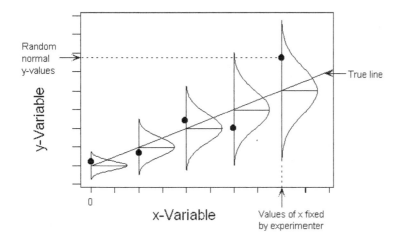

Fig. 6.7.1. The heteroscedastic model of regression.

with concentration. This probably happens also with shorter-range calibrations, but the change is small enough to escape detection, or small enough to ignore. The phenomenon is called heteroscedasticity, and it violates one of the assumptions on which the simple regression model is based. If heteroscedasticity is sufficiently marked, an adaptation of the model is required for best results. The model for this situation is shown in Fig. 6.7.1. The y-values (in calibration the analytical responses) are drawn at random from a distribution of which the standard deviation varies with the x-variable (concentration). (Compare this with Fig. 5.1.1.)

This adaptation, which takes account of the changes in variance, is called weighted regression. It works by loading each observation with a weight that is inversely proportional to the variance at the respective concentration. In that way, the regression 'takes more notice' of points with larger weights (smaller variances). The regression formulae tabulated below are similar to (and a generalisation of) those for simple regression. The important thing here is not to remember the equations, but to see the analogies between weighted and unweighted regression. Note that the weights are scaled so that the sum is equal to n, which simplifies the formulae somewhat. It should also be apparent that simple regression is a special case of weighted regression in which all of the weights are equal.

Statistic	Standard regression	Weighted regression
Data	$x_1, x_2, \ldots, x_i, \ldots, x_n$ $y_1, y_2, \ldots, y_i, \ldots, y_n$	$x_1, x_2, \ldots, x_i, \ldots, x_n$ $y_1, y_2, \ldots, y_i, \ldots, y_n$ $s_1, s_2, \ldots, s_i, \ldots, s_n$
Weight	$w_i = 1$	$w_i = n(1/s_i^2)/\sum_i (1/s_i^2)$, (i.e., $\sum_i w_i = n$)
Mean	$\bar{x} = \sum_i x_i/n$	$\bar{x}_w = \sum_i w_i x_i/n$
Slope of regression	$b = \dfrac{\sum_i (x_i - \bar{x})(y_i - \bar{y})}{\sum_i (x_i - \bar{x})^2}$	$b_w = \dfrac{\sum_i w_i (x_i - \bar{x}_w)(y_i - \bar{y}_w)}{\sum_i w_i (x_i - \bar{x}_w)^2}$
Intercept	$a = \bar{y} - b\bar{x}$	$a_w = \bar{y}_w - b_w \bar{x}_w$
Residual variance	$s_{y\mid x}^2 = \sum_i (y_i - \hat{y})^2/(n-2)$	$s_{y\mid x(w)}^2 = \sum_i w_i(y_i - \hat{y}_w)^2/(n-2)$
Variance of slope	$s_b^2 = s_{y\mid x}^2 / \sum_i (x_i - \bar{x})^2$	$s_{b(w)}^2 = s_{y\mid x(w)}^2 / \sum_i w_i (x_i - \bar{x}_w)^2$
Variance of intercept	$s_a^2 = s_b^2 \sum_i x_i^2/n$	$s_{a(w)}^2 = s_{b(w)}^2 \sum_i x_i^2/n$
Variance of evaluated x-value*	$s_{x_e}^2 = \dfrac{s_{y\mid x}^2}{b^2}$ $\times \left(\dfrac{1}{m} + \dfrac{1}{n} + \dfrac{(y' - \bar{y})^2}{b^2 \sum_i (x_i - \bar{x})^2} \right)$	$s_{x_e(w)}^2 = \dfrac{s_{y\mid x(w)}^2}{b_w^2}$ $\times \left(\dfrac{1}{w'} + \dfrac{1}{n} + \dfrac{(y' - \bar{y}_w)^2}{b_w^2 \sum_i w_i (x_i - \bar{x}_w)^2} \right)$

*This is the variance of unknown x-values x' calculated from the regression equation as $x' = (y' - a)/b$ by means of a measured response y', which is the mean of m measurements or has a corresponding weight w'.

The calculations of weighted regression are available in the usual statistical packages, so can be executed with ease. The only extra labour comprises the estimation of the weights. This may not be justified when the degree of heteroscedasticity is small. However, in long-range calibration the change in precision is often sufficient to have a deleterious impact on the resulting statistics. In that case, use of a weighted regression is recommended but, fortunately, even rough estimates of the weights improve the outcome substantially.

Weights can be estimated either from repeat measurements results or from a general experience of the performance of the analytical system. (For an example of the latter case, we might assume that the measurement standard deviation is 1% of the result except at zero concentration, where it is half of the detection limit.) In the example given below, weights are estimated from a small number of repeat measurements.

Further reading

- *'Why are we weighting?' (2007). AMC Technical Briefs. No. 27. Free download via www.rsc.org/amc*

6.8 Example of Weighted Regression — Calibration for ^{239}Pu by ICP–MS

Key points
— The ICPMS data for calibration of ^{239}Pu are markedly heteroscedastic.
— Weights were calculated by smoothing the initial estimates.
— Weighted regression gives statistics that are clearly more appropriate for the measurement of small concentrations.

The calibration data and calculation of the weights are shown in the table below (units are ng l^{-1}).

Concentration	R1	R2	R3	Raw SD	Smoothed SD	Variance	Weight
0	548	662	1141	315	231	53504	16.7015
2	6782	9661	9316	1572	1055	1112196	0.8035
4	15966	14067	17063	1516	1878	3526528	0.2534
6	25612	30337	26987	2430	2701	7296499	0.1225
8	30483	32143	35701	2666	3525	12422109	0.0719
10	42680	46291	35968	5239	4348	18903359	0.0473

Columns R1 to R3 show three repeat responses for each concentration. The next column shows the standard deviations calculated from the three responses. Such estimates are, of course, very variable and a better estimate might be obtained by smoothing them by simple regression, as in Fig. 6.8.1, or even by eye. The fitted values are shown in the following column. When conducting this smoothing it is important to check that all of the resulting estimates are reasonable and greater than zero. The variance is the square of the smoothed standard deviation and the weights are calculated from

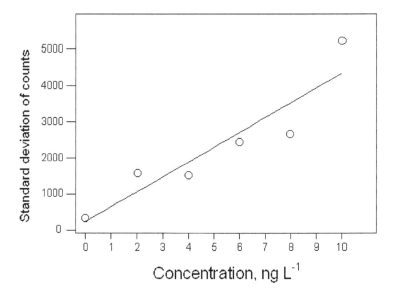

Fig. 6.8.1.   Smoothing the raw standard deviation estimates (circles) by regression (line).

$w_i = n(1/s_i^2)/\sum_i(1/s_i^2)$, so that the sum of the weights is 18 (the total number of measured responses).

The statistics from the weighted regression are as follows.

Predictor	Coefficient	Standard error	t	p
Intercept	767.8	145.8	5.27	0.000
Slope	4055.2	135.0	30.05	0.000
	$s_{y\vert x} = 142.0$	$R^2 = 98.3\%$		

Analysis of variance

Source of variation	Degrees of freedom	Sum of squares	Mean square	F	p
Regression	1	18208348	18208348	902.70	0.000
Residual error	16	322734	20171		
–Lack of fit	4	72606	18152	0.87	0.509
–Pure error	12	250128	20844		
Total	17	18531082			

Note the following.

- The intercept is significantly different from zero ($p < 0.0005$).
- As we have repeated measurements at each concentration, we can also conduct the test for lack of fit, and find that there is no significant lack of fit ($p = 0.509$).
- We can estimate a detection limit (§9.6) from $c_L = (3\mathrm{se}(a))/b = (3 \times 146)/4055 = 0.1\,\mathrm{ng\,l^{-1}}$.
- The regression line and the 95% confidence limits are shown in Fig. 6.8.2. The confidence interval is least near the origin of the graph.

We can see the beneficial effects of weighted regression in this instance by repeating the regression without weighting. The statistics are given below.

Predictor	Coefficient	Standard error	t	p
Intercept	559	1119	0.50	0.624
Slope	4126.1	184.8	22.32	0.000
	$s_{y\mid x} = 2679$	$R^2 = 96.9$		

Analysis of variance

Source of variation	Degrees of freedom	Sum of squares	Mean square	F	p
Regression	1	3575212011	3575212011	498.31	0.000
Residual error	16	114794611	7174663		
–Lack of fit	4	24146489	6036622	0.80	0.548
–Pure error	12	90648122	7554010		
Total	17	3690006622			

The most obvious differences from the weighted statistics are as follows.

- The standard error of the intercept is now incorrectly much greater (1119 instead of 146) and as a consequence we are tempted to make the incorrect inference that it is not significantly different from zero ($p = 0.624$).
- The apparent detection limit is now degraded to $(3 \times 1119)/4125 = 0.8\,\mathrm{ng\,l^{-1}}$, that is, magnified by a factor of about eight.

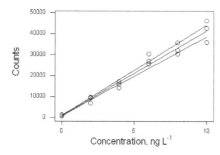   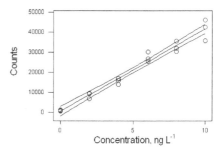

Fig. 6.8.2. Calibration data (circles) for Pu^{239} by ICP–MS, showing weighted regression line and its 95% confidence interval.

Fig. 6.8.3. Same calibration data as Fig. 6.8.2, with regression line and confidence intervals calculated by unweighted regression.

- The regression line is not much changed (Fig. 6.8.3) but the 95% confidence limit is much wider at low concentrations.

Therefore, for accurate work, especially at low concentrations, using weighted regression in calibration may be worth the minor degree of extra effort involved.

Further reading

- *'Why are we weighting?' (2007). AMC Technical Briefs No. 27. Free download via www.rsc.org/amc*

6.9   Non-linear Regression

Occasionally in analytical science we want to study the relationship between variables where the proposed model is non-linear in the coefficients (β). (This is 'non-linear' in the technical mathematical sense.) For instance, a model proposed for the variation in the uncertainty of measurement (u) in a particular method as a function of the concentration of the analyte (c) is as follows:

$$u = \sqrt{\beta_0^2 + \beta_1^2 c^2}.$$

This expression is not directly tractable by standard regression methods. In other words, if we minimise the sum of the squared residuals, the resulting normal equations cannot be solved for β_0, β_1 by straight algebraic methods. However, if we simply square the expression, we have

$$u^2 = \beta_0^2 + \beta_1^2 c^2,$$

and writing $\alpha_0 = \beta_0^2$, $\alpha_1 = \beta_1^2$ gives us

$$u^2 = \alpha_0 + \alpha_1 c^2,$$

which *is* linear in the coefficients and can be handled by simple regression by regarding u^2 as the dependent variable and c^2 as the independent variable.

Another example of this type is encountered in exploring how reproducibility standard deviation varies with concentration of the analyte in interlaboratory studies such as proficiency tests. We might suspect that the data could follow a generalised version of the Horwitz function (see §9.7) with unknown parameters, namely

$$\sigma_R = \beta_0 c^{\beta_1}.$$

Again this function cannot be tackled directly by regression. However, transforming the variables simply by taking logarithms gives us

$$\log \sigma_R = \log \beta_0 + \beta_1 \log c.$$

Now we can regress the dependent variable ($\log \sigma_R$) against the independent variable ($\log c$) to obtain estimates of the parameters $\alpha_0 = \log \beta_0$ and β_1. There is a slight difficulty in that the transformed error term may not be normally distributed, making the tests of significance non-exact, but this is seldom a serious objection. As usual we should look at the residual plots to ensure that the model is adequate.

However, there is another class of non-linear equations that cannot be treated by transformation. For example, there are theoretical reasons for expecting a calibration curve in ICP-AES to follow the pattern

$$r = \beta_0 + \beta_1 c - e^{\beta_2 + \beta_3 c}.$$

There is no transformation that will reduce this equation to a linear form. There are several ways of finding the estimates of the β_i, but all of these numerical methods depend on iterative procedures starting from initial approximations. Such methods are beyond the scope of these notes.

6.10   Example of Regression with Transformed Variables

> **Key points**
> — Reproducibility standard deviation was related to concentration of the
>   analyte (protein nitrogen in feed ingredients) using a non-linear model
>   analogous to the Horwitz function.
> — The data were logtransformed before regression.
> — No lack of fit was detected.

In a collaborative trial (method performance study) 14 different animal feed ingredients were subjected to the determination of the concentration c of nitrogen (as an indicator of protein content) by the Dumas method in the participating laboratories. The standard deviation of reproducibility σ_R was calculated for each material. The investigator wished to see whether the results conformed to the Horwitz function (see §9.7),

$$\sigma_R = 0.02 c^{0.8495},$$

or, transforming to logarithms base 10,

$$\log \sigma_R = \log 0.02 + 0.8495 \log c = -1.699 + 0.8495 \log c.$$

The primary statistics obtained were as follows. (All the data are expressed as mass fractions, as required by the Horwitz function so, for example, $0.0946 \equiv 9.46\%$ mass fraction.)

c	s_R	$\log c$	$\log s_R$
0.0946	0.0011	−1.02411	−2.95861
0.1197	0.002488	−0.92191	−2.60415
0.0271	0.0007	−1.56703	−3.1549
0.1867	0.003062	−0.72886	−2.51399
0.0188	0.000328	−1.72584	−3.48413
0.0827	0.001444	−1.08249	−2.84043
0.0141	0.000404	−1.85078	−3.39362
0.0168	0.00046	−1.77469	−3.33724
0.0267	0.000572	−1.57349	−3.2426
0.0443	0.001022	−1.3536	−2.99055
0.1049	0.001498	−0.97922	−2.82449
0.0111	0.000284	−1.95468	−3.54668
0.0636	0.001306	−1.19654	−2.88406
0.0436	0.000814	−1.36051	−3.08938

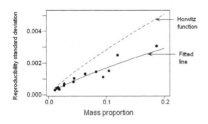

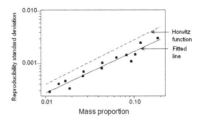

Fig. 6.10.1. Nitrogen data (solid circles) with fitted line (solid) and Horwitz function (dashed).

Fig. 6.10.2. Nitrogen data (solid circles) with fitted line (solid) and Horwitz function (dashed). Same data and functions as Fig. 6.10.1 but on logarithmic axes.

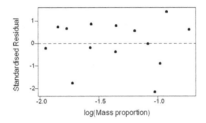

Fig. 6.10.3. Residuals from the fit to the logtransformed nitrogen data.

Regression of $\log s_R$ on $\log c$ gave the following output.

Predictor	Coefficient	Standard error	t	p
Intercept	−1.97935	0.08363	−23.67	0.000
Slope	0.79366	0.05915	13.42	0.000
	$s_{y|x} = 0.0824$	$R^2 = 93.8$		

The regression therefore tells us that

$$\log s_R = -1.979 + 0.794 \log c.$$

As $10^{-1.979} = 0.0105$, transforming back to mass fractions gives us

$$s_R = 0.0105 c^{0.794}.$$

The exponent found is somewhat lower than that of the Horwitz function, although probably not significantly so. The coefficient of 0.0105 is considerably lower than the Horwitz value, however, showing that the determination is more precise than predicted. These features can be seen in Figs. 6.10.1 and 6.10.2, together with the Horwitz function. The residual plot (Fig. 6.10.3) suggests a reasonable fit — the residuals look a bit skewed — but the deviation from normality is not significant at the 95% level of confidence.

Chapter 7

Additional Statistical Topics

This chapter covers a number of practical topics related to the normal distribution and deviations from it that are relevant to the work of analytical chemists. Outlier tests are covered, but the superiority of robust methods in this area is emphasised.

7.1 Control Charts

Key points
— Control charts are used to check that a system is operating 'in statistical control'.
— Shewhart charts traditionally have 'warning limits' at $\mu \pm 2\sigma$ and 'action limits' at $\mu \pm 3\sigma$.
— A system is regarded as out of control on the basis of a result outside the action limits.

An analytical system where all the factors that affect the magnitude of errors are kept constant is said to be in 'statistical control'. Under those conditions it would be reasonable to assume that results obtained by repeated analysis of a single test material would resemble independent random values taken from a normal distribution $N(\mu, \sigma^2)$. Thus we would expect in the long term about 95% of results to fall within a range of $\mu \pm 2\sigma$ and about 99.7% to fall within the range $\mu \pm 3\sigma$. A result obtained outside the latter range would be so unusual under the assumption of statistical control that it is conventionally taken to indicate that the assumption is invalidated, i.e., that conditions determining the size of errors have changed, and the analytical system is behaving in a new and unacceptable fashion. Either a new value of μ prevails, or a larger value of σ, perhaps because of instrument malfunction or a new batch of reagents, or failure on the part of the analyst to observe some aspects of the operating procedure. This requires some investigation of the system and remediation where necessary to restore the initial conditions.

A convenient way to monitor an analytical system in this way is via a control chart. This is based on the results obtained by the analysis of one or more special test materials that have been homogenised and tested for stability. These 'control materials' must be typical of the material under routine test and are analysed exactly as if they were normal samples in every run of the analytical system. The results are plotted on a chart that shows the results as a function of run number. The chart conventionally has lines at μ, $\mu \pm 2\sigma$ ('Warning limits'), and $\mu \pm 3\sigma$ ('Action limits') (Fig. 7.1.1). This type of chart is called a Shewhart chart.

Under statistical control, a result is very unlikely to fall outside the action limits. Such a point is taken to show that the system is 'out of control', requiring the results obtained in that run to be regarded as suspect, and the analytical system to be halted until statistical control has been restored. Some other occurrences are about equally unlikely, and are also taken to indicate out-of-control conditions, namely: (a) two successive results outside the warning limits; or (b) nine consecutive results on the same side of the mean line.

There are many different kinds of control chart with differing capabilities. The Shewhart chart is good for detecting abrupt changes in the

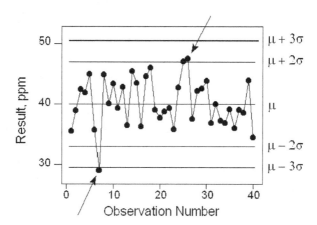

Fig. 7.1.1. Shewhart control chart showing results for a control material in successive runs. The arrows show the system going out of control at Run 7 and Run 26.

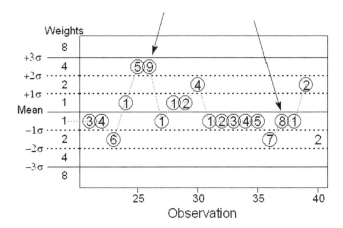

Fig. 7.1.2. Zone chart showing part of the same data as Fig. 7.1.1. Symbol positions (numbered circles) show the zone of the current result. Numbers show the accumulated score. The arrows show Run 26 to be out of control as before, but also shows Run 37 to be out of control.

analytical system. Other control charts are better at detecting smaller changes or a drift. The Cusum chart is one such. The Zone chart (Fig. 7.1.2) has roughly the combined capabilities of the Shewhart chart and the Cusum chart and is simple to plot and interpret. In this chart, results are converted into scores that depend on which zone of the chart the results falls into. These scores are labelled weights in the figure. With each successive result the scores are aggregated. If a new result falls on the opposite side of the mean from the previous one, the total score is rest to zero before the new score is aggregated. If the aggregate score gets to eight or more, the system is deemed out of control. The Zone chart detects all of the out-of-control conditions that the Shewhart chart does but also detects the smaller change that results in the score of eight at Run 37.

Further reading

- *Thompson, M. and Wood, R. (1995). Harmonised Guidelines for Internal Quality Control in Analytical Chemistry Laboratories, Pure Appl Chem, 67, pp. 649–666.*
- *'The J-chart: a simple plot that combines the capabilities of Shewhart and Cusum charts, for use in analytical quality control'. (2003). AMC Technical Briefs No. 12. Free download via www.rsc.org/amc.*
- *'Internal quality control in routine analysis'. (2010). AMC Technical Briefs No. 46. Free download via www.rsc.org/amc.*

7.2   Suspect Results and Outliers

Key points
— Analytical datasets often contain suspect values that seem inconsistent with the majority.
— Outliers can have a large influence on classical statistics, especially standard deviations.
— It is often difficult to identify suspect values visually as outliers.
— Deleting identified outliers before calculating statistics needs an informed judgement.

While a set of repeated analytical results will often broadly resemble a normal distribution, it is not uncommon to find that a small proportion of the results are discrepant, that is sufficiently different from the rest of the results to make the analyst suspect that they are the outcome of a large uncontrolled variation (i.e., a mistake) in procedure. Data given below and shown in Fig. 7.2.1 can be taken as a typical example.

$$15.1 \quad 24.9 \quad 26.7 \quad 27.1 \quad 28.4 \quad 31.1$$

Outliers have a large effect on classical statistics (especially the standard deviation), which could invalidate decisions depending on probabilities. This effect can be seen by comparing Figs. 7.2.2 and 7.2.3. Where there is a documented mistake that accounts for the suspect value, it can be corrected or deleted from the dataset without any question. When there is no such explanation, analysts differ about whether deletion is justified. Those against deletion argue that the discrepant result is still part of the analytical system so should be retained if the summary statistics are to be fully descriptive and suitable for prediction of future results from the analytical system. In any event, deletion seems like an unhealthy subjectivity creeping into the science. Other scientists maintain that deletion is appropriate when

Results for total aflatoxins, ppb

Fig. 7.2.1.   Results of a determination repeated by five analysts. The result at 15.1 is suspected of being an outlier.

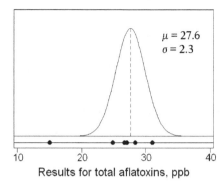

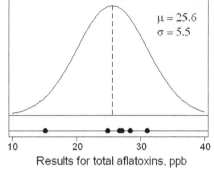

Results for total aflatoxins, ppb

Results for total aflatoxins, ppb

Fig. 7.2.2. The normal distribution modelling the data if the suspect value is excluded. Most of the data are modelled well.

Fig. 7.2.3. The normal distribution modelling the data if the suspect value is included. The mean seems biased low and the standard deviation is inflated. All of the data are modelled poorly.

the suspect result is clearly discrepant, so that summary statistics accurately represent the behaviour of the great majority of the results. These scientists have to bear in mind that outliers may occur in the future, although they will have no indication of their probability or magnitude. Figures 7.2.2 and 7.2.3 illustrate these points. Both of these arguments have some virtue in particular circumstances, but the crucial decision is whether the suspect point is really discrepant or is simply a slightly unusual selection of results from a normal distribution. Our visual judgement of this is notoriously poor when, as typically in analytical science, there are only a few data available.

Statistical tests for outliers abound, but they tend to suffer from various defects. The simple version of Dixon's test (§7.3), for example, will not give a sensible outcome if there are two outliers present, at either end of the distribution. A better way of handling suspect data is the use of robust statistics (see §7.5, 7.6).

Notes and further reading

- *'Rogues and suspects: how to tackle outliers'. (April 2009). AMC Technical Briefs No. 39. Free download via www.rsc.org/amc.*
- *The dataset can be found in the file named* **Suspect.**

7.3   Dixon's Test for Outliers

Key points

— Dixon's test compares the distance from the suspect value to its nearest value with the range of the data.
— The simple version of Dixon's test is foiled by the presence of a second outlier.

A simple test of a suspect value is Dixon's Q test. The test statistic Q is the distance from the suspect point to its nearest neighbour divided by the range of the data: in terms of Fig. 7.3.1 we have $Q = A/B$. For our example data we have

$$Q = A/B = (24.9 - 15.1)/(31.1 - 15.1) = 0.61.$$

The probability of a value of Q exceeding 0.61 arising from random samples of six observations from a normal distribution is about 0.07. This is almost small enough to reject the null hypothesis (no outlier) at 95% confidence so we could reasonably treat 15.1 as an outlier. (In this case a Q value of greater than 0.63 would be required to provide 95% confidence.)

A problem with this simple test is that it can be foiled by the presence of a second outlier at either end of the range. Suppose in addition to the previous results there was an extra value at 37.1 ppb (Fig. 7.3.2). The test statistic would then be

$$Q = A'/B' = ((24.9 - 15.1)/(37.1 - 15.1)) = 0.45.$$

A value as high as 0.45 would arise by chance with a probability as high as 0.17, so we should now be unwilling to regard the low value as an outlier. Indeed, it is *not* an outlier even though it is the same distance as before from the closest value! A similar problem would arise if the extra value were on

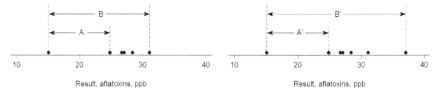

Fig. 7.3.1.   Dixon's test $Q = A/B$ applied to the aflatoxin data.   Fig. 7.3.2.   Dixon's test applied to an extended dataset.

the same extreme as the original suspect value. There are modified versions of Dixon's test that can be applied to this new situation, but these and similar problems affect many outlier tests. A more sophisticated treatment of suspect values is preferable (§7.5, 7.6).

Notes and further reading

- *'Rogues and suspects: how to tackle outliers'. (April 2009). AMC Technical Briefs No. 39. Free download via www.rsc.org/amc.*
- *Critical values of Q can be found in many statistical texts.*
- *The dataset is in file named **Suspect**.*

7.4 The Grubbs Test

Key points

— Grubbs test in the simplest form tests for an outlier by calculating the difference between the largest (or smallest) value and the mean of a dataset with respect to the standard deviation.
— It can be expanded to test for multiple outliers.

This is a more sophisticated test for outliers than Dixon's test. It is used to detect outliers in a dataset by testing for one outlier at a time. Any outlier which is detected is deleted from the data and the test is repeated until no outliers are detected. However, multiple iterations change the probabilities of detection, and the test should not be used for sample sizes of six or less since it frequently tags most of the points as outliers. The basic assumption underlying the Grubbs test is that, outliers aside, the data are normally distributed. The null hypothesis is that there are no outliers in the dataset.

The test statistic G is calculated for each result x_i from the sample mean $\bar{x}$ and standard deviation s as

$$G = \max |x_i - \bar{x}|/s.$$

This statistic calculates the value with the largest absolute deviation from the sample mean in units of the sample standard deviation. This form of the Grubbs test is therefore a two-tailed test. There are other forms of the test including one-tailed versions.

As with Dixon's test, the aflatoxin data can be used as an example.

$$15.1 \quad 24.9 \quad 26.7 \quad 27.1 \quad 28.4 \quad 31.1$$

We carry out a two-tailed test as above under the null hypothesis that there is no outlier in the data set. The test statistic is calculated as

$$G = |15.1 - 25.55|/5.5186 = 1.894.$$

This is compared with the 95% critical value for $n = 6$ of 1.933. As the calculated value of G is less than the critical value, the null hypothesis is not rejected at 95% confidence and the result 15.1 is not identified as an outlier. However, G is high enough at least to warrant checking the calculations leading to the result.

Notes and further reading

- *'Grubbs' is the name of the originator of this test, and is not a possessive case.*
- *'Rogues and suspects: how to tackle outliers'. (April 2009). AMC Technical Briefs No. 39. Free download via www.rsc.org/amc.*
- *Tables of critical levels of the test statistic can be found in some statistical texts. A brief table for the two-tailed test is given below, with N the number of observations.*

N	$G_{95\%}$	$G_{99\%}$
6	1.933	1.993
7	2.081	2.171
8	2.201	2.316
9	2.3	2.438
10	2.383	2.542
11	2.455	2.631
12	2.518	2.709
13	2.574	2.778
14	2.624	2.84
15	2.669	2.895
16	2.71	2.945
17	2.748	2.991
18	2.782	3.032
19	2.814	3.071
20	2.843	3.106

- *In the absence of tables, critical levels can be calculated from the equation* $\dfrac{N-1}{\sqrt{N}}\sqrt{\dfrac{t^2_{(\alpha/2N),\,N-2}}{N-2+t^2_{(\alpha/2N),\,N-2}}}$ *with* $t_{(\alpha/2N),\,N-2}$ *denoting the critical value of the t distribution with $(N-2)$ degrees of freedom and a significance level of $\alpha/2N$. (α is the overall significance level so, for $G_{95\%}, \alpha = 0.05$.)*

7.5 Robust Statistics — MAD Method

> **Key points**
> — Robust methods reduce the influence of outlying results and provide statistics that describe the distribution of the central or 'good' part of the data.
> — The methods are applicable to data that seem to be unimodal and roughly symmetrically distributed.
> — Robust statistics must be used with care for prediction.
> — The MAD method is a quick way of calculating robust mean and standard deviation, without requiring any decisions about rejecting outliers.

The use of robust statistics enables us to circumvent the sometimes contentious issue of the deletion of outliers and provides perhaps the best method of identifying them. Robust methods of estimating statistics such the mean and standard deviation (and many others) reduce the influence of outlying results and heavy tails in distributions. Robust statistics requires the original data to be similar to normal (i.e., roughly symmetrical and unimodal) but with a small proportion of outliers or heavy tails. It cannot be used meaningfully to describe strongly skewed or multimodal datasets. Robust statistics must be used with care in prediction. They will not enable the user to predict the probability or likely magnitude of outliers.

There are a number of robust methods in use. One of the simplest is the MAD (Median Absolute Difference) method. Suppose we have replicated results as follows:

$$145 \quad 130 \quad 157 \quad 153 \quad 183 \quad 148 \quad 151 \quad 143 \quad 147 \quad 163.$$

Putting these in ascending order we have:

$$130 \quad 143 \quad 145 \quad 147 \quad \mathbf{148} \quad \mathbf{151} \quad 153 \quad 157 \quad 163 \quad 183.$$

We need the median of these results, that is, the central value. In this instance the median is the mean of 148 and 151, namely 149.5. This median is a robust estimate of the mean, that is, $\hat{\mu} = 149.5$. It is unaffected by making the extreme values more extreme, for instance, by changing the size of the lowest value to any value lower than 148, and/or the highest value to any value greater than 151.

The next stage is to subtract the median of the data from each value and ignore the sign of the difference, giving the absolute differences (in the same order as immediately above):

$$19.5 \quad 6.5 \quad 4.5 \quad 2.5 \quad 1.5 \quad 1.5 \quad 3.5 \quad 7.5 \quad 13.5 \quad 33.5$$

If we sort these absolute differences into increasing order, we have:

$$1.5 \quad 1.5 \quad 2.5 \quad 3.5 \quad \mathbf{4.5} \quad \mathbf{6.5} \quad 7.5 \quad 13.5 \quad 19.5 \quad 33.5.$$

The median of these results (the median absolute difference) is the mean of 4.5 and 6.5, namely 5.5. This also is unaffected by the magnitude of the extreme values. We multiply this median by the factor 1.4825, which is derived from the properties of the normal distribution. The product is the robust estimate of the standard deviation, namely $\hat{\sigma} = 5.5 \times 1.4825 = 8.2$ (to two significant figures). (Note: the robust statistics are designated $\hat{\mu}, \hat{\sigma}$ to distinguish them from the classical estimators $\bar{x}, s$, which are used only as defined in §2.3.)

The robust estimates are thus $\hat{\mu} = 149.5$, $\hat{\sigma} = 8.2$.

The MAD method is quick and has a negligible deleterious effect on the statistics if the dataset does not include outliers. It therefore can be used in emergencies (i.e., when there isn't a calculator handy). There are somewhat better ways of estimating robust means and standard deviations, but they all require special programs to do the calculations.

Notes and further reading

- *A factor of 1.5 applied to the MAD (instead of 1.4825) is accurate enough for all analytical applications.*
- *'Rogues and suspects: how to tackle outliers'. (April 2009). AMC Technical Briefs No. 39. Free download from www.rsc.org/amc.*
- *Rousseeuw, P.J. (1991). Tutorial to Robust Statistics, J Chemomet, **5**, pp. 1–20. This paper can be downloaded gratis from ftp://ftp.win.ua.ac.be/pub/preprints/91/Tutrob91.pdf.*
- *The dataset used in this example is named is **Outlier**.*

7.6   Robust Statistics — Huber's H15 Method

Key points

— Huber's H15 method is a procedure for calculating robust mean and standard deviation.

— It is an iterative procedure, starting with initial estimates of the statistics.

— At each iteration, the data are 'winsorised' (modified) by using the current values of the statistics.

— Comparing data with a robust fit is one of the best ways of identifying suspect values.

Several methods for robustifying statistical estimates depend on down-weighting extreme values in some way, to give them less influence on the outcome of the calculation. Huber's H15 method is one such that has been widely used in the analytical community. Like most of these methods, it relies on taking initial rough estimates of the statistics $(\hat{\mu}_0, \hat{\sigma}_0)$ and refining them by an iterated procedure.

Using the data from §7.5 we subject it to a process called 'Winsorisation'. This involves replacing original datapoints x falling outside the range $\hat{\mu}_0 \pm k\hat{\sigma}_0$ with the actual range limits. This creates pseudo-values $\tilde{x}$ thus:

$$\tilde{x}_1 = \begin{cases} \hat{\mu}_0 + k\hat{\sigma}_0, & \text{if } x > \hat{\mu}_0 + k\hat{\sigma}_0 \\ \hat{\mu}_0 - k\hat{\sigma}_0, & \text{if } x < \hat{\mu}_0 - k\hat{\sigma}_0 \\ x, & \text{if } \hat{\mu}_0 - k\hat{\sigma}_0 < x < \hat{\mu}_0 + k\hat{\sigma}_0 \end{cases}.$$

The first revised estimates of the statistics are then given by $\hat{\mu}_1 = \text{mean}(\tilde{x}_1)$ and $\hat{\sigma}_1 = \text{sd}(\tilde{x}_1)/\theta$. For a moderate proportion of outlying results (and most analytical applications) k can be set at 1.5. The corresponding value of θ, derived from the properties of the normal distribution, is 0.882. The process is then repeated using $\hat{\mu}_1$, $\hat{\sigma}_1$ to winsorise the data and calculate the improved estimates $\hat{\mu}_2$, $\hat{\sigma}_2$ in the same manner, and so on until a sufficient degree of convergence is obtained. Convergence is slow so a computer is required.

The table below shows the application of this to the previously-used suspect data (row x_0) starting with the MAD estimates $\hat{\mu}_0 = 149.5$, $\hat{\sigma}_0 = 8.15$. The first replacement limits are $\hat{\mu}_0 \pm k\hat{\sigma}_0 = (161.73, 137.27)$, so any value less that 137.27 becomes 137.27 and any value greater than

161.73 becomes 161.73. Thus in the first Winsorisation (row $\tilde{x}_1$) three values (boldface) are replaced, producing estimates of $\hat{\mu}_1 = 150.47$, $\hat{\sigma}_1 = 9.11$ to two decimal places.

Data										$\hat{\mu}_{rob}$	$\hat{\sigma}_{rob}$
x_0	130.00	143	145	147	148	151	153	157	163.00	183.00	149.50	8.15
$\tilde{x}_1$	**137.27**	143	145	147	148	151	153	157	**161.73**	**161.73**	150.47	9.11
$\tilde{x}_2$	**136.81**	143	145	147	148	151	153	157	163.00	**164.14**	150.80	9.87
$\tilde{x}_3$	**135.98**	143	145	147	148	151	153	157	163.00	**165.61**	150.86	10.33
$\tilde{x}_4$	**135.36**	143	145	147	148	151	153	157	163.00	**166.36**	150.87	10.62
$\vdots$											$\vdots$	$\vdots$
$\tilde{x}_{17}$	**134.21**	143	145	147	148	151	153	157	163.00	**167.54**	150.88	11.11

In subsequent iterations only two values are replaced. The results have stabilised sufficiently by the 17th iteration, giving final estimates of $\hat{\mu}_{rob} = 150.88$, $\hat{\sigma}_{rob} = 11.11$. These can be compared with the classical statistics, for the complete data ($\bar{x} = 152.0$, $s = 14.0$) and for the data with the suspect value deleted ($\bar{x}' = 148.0$, $s' = 9.33$). Simply deleting the suspect value gives a standard deviation that is too low.

Robust statistics is probably one of the best ways of identifying outliers. If we pseudo-standardise the dataset as $z = (x - \hat{\mu}_{rob})/\hat{\sigma}_{rob}$, the 'good' results should resemble a sample from the standard normal distribution. Any results with a magnitude greater than about 2.5 can therefore be regarded as at least suspect, if not outlying. If we apply this transformation to our example data (in increasing order) we obtain:

$$z \quad -1.9 \quad -0.7 \quad -0.5 \quad -0.3 \quad -0.3 \quad 0.0 \quad 0.2 \quad 0.6 \quad 1.1 \quad \mathbf{2.9}.$$

The value of 2.9 suggests that the original result of 183 is suspect and that its provenance should be investigated further at least.

Notes and further reading

- 'Rogues and suspects: how to tackle outliers'. (April 2009). AMC Technical Briefs No. 39. Free download from www.rsc.org/amc.
- Analytical Methods Committee. (1989). Robust Statistics — How Not to Reject Outliers. Part 1. Basic Concepts, Analyst, **114**, pp. 1693–1697.
- There is Excel software for conducting the H15 method in AMC Software free download via www.rsc.org/amc.

- *Rousseeuw, P.J. (1991). Tutorial to Robust Statistics, J Chemomet, **5**, pp. 1–20. This paper can be downloaded gratis from ftp://ftp.win.ua.ac. be/pub/preprints/91/Tutrob91.pdf.*
- *The dataset used in this example is named is **Outlier**.*

7.7 Lognormal Distributions

Key points
— A variable x is lognormally distributed if log x is normally distributed.
— The shape of a lognormal distribution depends on its relative standard deviation.
— Lognormal distributions of error are rare in chemical measurement.
— Some analytical circumstances give rise to distributions with a quasi-lognormal distribution.
— Logtransformation sometimes can be safely used to stabilise variance before regression or ANOVA.

A variable x is lognormally distributed if log x is normally distributed. Figure 7.7.1 shows a lognormal distribution with a mean of two and a standard deviation of one, that is, with a relative standard deviation (RSD) of 50%. It has zero density (height) when x is zero, and a positive skew. All lognormal distributions (but many others) have these two properties. A plot of density against log x (Fig. 7.7.2) has the familiar shape of the normal distribution. The shape (degree of skewness) of the lognormal distribution depends on the RSD. For instance, a distribution with an RSD of 10%

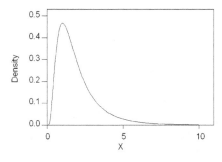

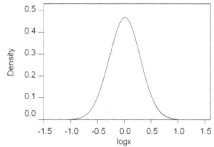

Fig. 7.7.1. A lognormal distribution with a relative standard deviation of 50%.

Fig. 7.7.2. The same distribution plotted against $\log_{10} x$.

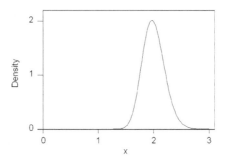

Fig. 7.7.3. Lognormal distribution with an RSD of 10%, showing little visible sign of asymmetry.

(Fig. 7.7.3) or lower is very different from one with an RSD of 50%, and hard to distinguish visually from a normal distribution.

A lognormal distribution in measurement implies that the errors are multiplicative. The physical circumstances of a chemical measurement rarely give rise to results that genuinely follow that rule. However, in one currently-important type of measurement — the determination of specific DNA sequences from genetically-modified food by the real-time polymerase chain reaction (PCR) — that circumstance is approximately realised. Because the fundamental procedure of PCR is multiplicative, the errors tend to follow the same pattern. So we might find the 95% confidence limits of repeated measurements of (say) 0.5 and 2.0 times the mean concentration. For this particular measurement a logtransformation of the results has been found to be justified and helpful.

In nearly all other types of chemical measurement, however, these conditions do not apply and logtransformation should be used with due caution. This is important to remember because appearances can be deceptive at concentrations near the detection limit, where the RSD is high — by definition greater than about 30%. Repeat results at low concentrations may sometimes appear to be similar to the lognormal because they have been censored or truncated at zero. The confusion arises because true concentrations cannot be below zero. However, the results of measurements are not true concentrations — they include errors — and they can and sometimes do fall below zero. Some analysts are uncomfortable with this apparent conflict and as a consequence do not record negative results. Despite the appearance of censored results, logtransformation will be misleading in this situation.

Another mathematical operation that gives rise to a skewed distribution (and causes the same type of confusion) is the division of one imprecise variable by another. That might happen in the correction of raw analytical results for recovery. It should not cause noticeable skewness except near the detection limit. In that region, and in combination with censoring results at zero, the resulting distribution could on casual inspection be taken as lognormal by the unwary.

Analytical chemists sometimes encounter other datasets where a variable is genuinely strictly positive and skewed. Concentrations of trace analytes in more-or-less random collections of natural materials (e.g., copper in sediments [§1.3]) usually have that property and sometimes approach lognormal. Again, samples taken from a contaminated site or area may show a similar distribution if the contamination is patchy. Log-transformation is sometimes useful in the characterisation of such data but, as always, should be used with due caution.

There are, however, situations where logtransformation can be helpful, and that is in regression and analysis of variance where the precision of the response varies widely but its RSD can be taken as constant. Of course, weighted regression will serve if the weights can be estimated, but often the information is not available. Logtransformation will stabilise the variance of the response in that situation, because different variables with the same RSD all have the same absolute standard deviation when logtransformed. For example, an RSD of 10% becomes a constant SD of 0.045 under logtransformation, regardless of the concentration. The trans-formed data will still have a distribution close to normal (unless the RSD is much higher than 10%), so the usual assumptions of simple regression can be made.

7.8   Rounding

> **Key point**
> — Round the standard deviation to two significant figures and round the mean to the same number of decimal points (or trailing zeros).

Modern computers use a large number of significant figures in calculations, and applications often provide statistics with an excessive number of sig-nificant figures. For example, a computer might tell us that the statistics for the data in §7.2 are: mean 25.5500; and standard deviation 5.5186. Such data need to be rounded before reporting, because it is obvious that the fourth decimal place (at least) is quite meaningless.

We are normally taught to retain only the first figure that is uncertain. But it is important not to overdo the rounding, which can destroy useful information. The commonly-used rule is sometimes too rigorous. This may be important if the results are to be used in further statistical operations. There is a simple rule for appropriate rounding when we report such data: round the standard deviation to two significant figures and round the mean to the same number of decimal points (or trailing zeros). The rationale here is that estimated standard deviations are hardly ever more accurate than that. Applying this rule to the statistics above gives us: mean 25.6; and standard deviation 5.5. This rule nearly always leaves a generous number of significant figures.

7.9   Non-parametric Statistics

Key points
— Non-parametric tests require less stringent assumptions than parametric tests. The assumption of normality is not required.
— Many parametric tests have non-parametric equivalents.

Non-parametric tests are sometimes called distribution-free statistics because they do not depend on the data being drawn from normal distributions. More generally, non-parametric tests require less restrictive assumptions about the data. Many statistical tests have non-parametric equivalents. An important reason for using these tests is that they allow the analysis of rank data. Despite these useful features, and their widespread use in the social sciences, non-parametric tests are seldom used by analytical chemists because parametric methods are more usually more powerful, and the normal distribution is often a reasonable assumption in physical measurement.

The most commonly used non-parametric test in analytical chemistry is perhaps the Mann–Whitney U test, which is explained in detail below. The Mann–Whitney test is the non-parametric equivalent of the two-sample t-test for comparing the central tendencies of two independent datasets. For a two-tailed test the null and alternative hypotheses are as shown here.

One tailed tests can also be carried out. For datasets P and Q, the null and alternative hypotheses are:

$$H_0 : \text{Median(P)} = \text{Median(Q)}$$
$$H_A : \text{Median(P)} \neq \text{Median(Q)}$$

The test statistic is obtained by calculating the lesser of U_P and U_Q, where

$$U_P = nm + \frac{m(m+1)}{2} - S_P,$$
$$U_Q = nm + \frac{n(n+1)}{2} - S_Q,$$

and data samples of size m (the larger set P) and n (the smaller set Q) are pooled. S_P, S_Q are the sums of the pooled ranks for the respective datasets.

The test is based only on the following assumptions: the datasets are independent random samples from the respective populations and the measurement scale is at least ordinal. A confidence interval for the difference between the population medians can be estimated with the further assumption that the two population distribution functions are identical apart from a possible difference in location.

An example is shown using the wheat flour data from §1.3.6, in which two methods are used to test for nitrogen in a sample of wheat flour. The ranked data (that is, tabulated in increasing order) are shown below along with the associated method (K = Kjeldahl, D = Dumas). Note that tied results (e.g., 2.92, 2.92) are each given the mean rank (1.5).

Result %	Rank	Result %	Rank
2.92 (K)	1.5	3.02 (K)	9.5
2.92 (K)	1.5	3.04 (D)	11.5
2.98 (K)	3.5	3.04 (D)	11.5
2.98 (D)	3.5	3.05 (K)	13.5
3.00 (K)	5	3.05 (D)	13.5
3.01 (K)	7	3.07 (K)	15
3.01 (K)	7	3.08 (D)	16.5
3.01 (D)	7	3.08 (D)	16.5
3.02 (K)	9.5	3.12 (D)	18

For the Kjeldahl (larger) dataset, $m = 10$ and the sum of ranks is $S_K = 73$.

For the Dumas (smaller) dataset, $n = 8$ and the sum of ranks is $S_D = 98$.

From this we have $U_K = 80 + 55 - 73 = 62$, and $U_D = 80 + 36 - 98 = 18$, so the value of the test statistic $U = 18$ is the lesser of these.

For the sample sizes in this example, the critical value for 95% confidence is 17. As $18 > 17$, the null hypothesis is rejected and the medians are significantly different. Most statistical software will provide this information plus a p-value, and confidence limits for the difference between the medians.

Notes and further reading

- *There are a number of other tests that are counterparts of a parametric test. These include the Wilcoxon Matched Pairs Signed Ranks test, which is the equivalent of the paired t-test, and the Kruskal–Wallis test, which is a method for comparing several independent random samples and which can be used as a non-parametric alternative to the one way ANOVA.*
- *Further details and tables of critical values can be found in standard reference books.*
- *Most statistical software packages provide all of the common non-parametric tests.*

7.10 Testing for Specific Distributions — the Kolmogorov–Smirnov One-Sample Test

---

Key points

— The Kolmogorov–Smirnov test is a non-parametric test used to test whether or not a single sample of data is consistent with a specified distribution function.

— The data values are ordered and compared with the equivalent value from the distribution.

The Kolmogorov–Smirnov statistic quantifies a distance between the empirical cumulative distribution function of the sample and the cumulative distribution function of the hypothesised distribution, often the normal distribution. A graphical representation is show in Fig. 7.10.1.

To calculate the test statistic it is necessary to calculate the values of
$F(x)$: the empirical cumulative density function
$G(x)$: the cdf from the hypothesised distribution. In this example it is the normal distribution.

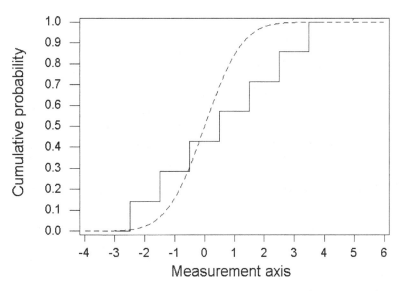

Fig. 7.10.1. Cumulative distribution functions of a test dataset (step function) and a normal distribution (dashed line) with mean zero and unit standard deviation.

Observed data	F(x)	G(x)	F(x) − G(x)
−3	0	0.0010	−0.0010
−2	0.1429	0.0183	0.1246
−1	0.2857	0.1379	0.1478
0	0.4286	0.500	−0.0714
1	0.5714	0.8414	−0.27
2	0.7143	0.9773	−0.236
3	0.8571	0.9987	−0.1416
4	1.0	1.0	0

The test statistic is:

$$D = \text{Max}(|F(x) - G(x)|)$$

$$D = 0.27$$

The hypotheses for the Kolmogorov–Smirnov test are defined as

H_0: The data follow a specified distribution
H_A: The data do not follow the specified distribution

The null hypothesis is rejected if the test statistic, D, is greater than the critical value which is provided by most statistical software. Tables can be found in textbooks and online.

For $n = 8$ the critical value $= 0.457$ at the 95% level. As $0.27 < 0.457$ then the null hypothesis cannot be rejected.

Assumptions:

• It only applies to samples from continuous distributions.
• The distribution must be fully specified. That is, if location, scale and shape parameters are estimated from the data, the critical region of the Kolmogorov–Smirnov test is no longer valid. It typically must be determined by simulation.

The distribution specified in the null hypothesis is often the normal distribution. Hence the Kolmogorov–Smirnov test is often used as a test for normality. Here the test compares the cumulative distribution of the data with the expected cumulative normal distribution, with the test statistic being based on the largest discrepancy. Other tests for normality also depend on the comparison of cumulative distributions.

Testing for normality should be treated with caution. If a dataset fails the null hypothesis it is not always the case that parametric tests cannot be applied. Consider the following points.

• Small samples almost always pass a normality test. Normality tests have little power to tell whether or not a small sample of data comes from a normal distribution.
• With large samples, minor deviations from normality may be flagged as statistically significant, even though small deviations from a normal distribution will not affect the results of a t-test or ANOVA.

7.11 Statistical Power and the Planning of Experiments

Key points
— The power of a statistical test is the probability of rejecting a null hypothesis when it is in fact false.
— Power calculations provide a way of checking whether a proposed experiment is capable of delivering a useful result at a minimal cost.

The outcome of a statistical test stems from a balance between various circumstances, namely the magnitude of the effect being tested, the precision of the measurements and the number of measurements made. (An effect is the deviation of the test statistic from H_0.) Suppose the outcome was 'not significant'. It might well be that the opposite outcome would have been recorded if the analytical method had been more precise or more repeat measurements made. Alternatively, an effect might be significant but of a magnitude that is of no importance in the context of the test. It is worthwhile considering this balance before any measurements are undertaken: there is no point in undertaking an experiment that is unlikely to provide a useful outcome. Equally, as measurements cost money, and higher precision costs more than lower, it is important to commit the least resource that will provide a useful outcome. These considerations come under the heading of statistical power.[1] It is therefore very good practice to estimate the power of a proposed experiment before it is undertaken.

Section 1.11 described a critical level of probability that we regard as convincing for the particular inference that we wish to make. In statistical power terminology, this critical probability is referred to as α. A 'Type I' error occurs when a true null hypothesis is rejected. The probability of a Type I error is equal to α.

A Type II error occurs when a false null hypothesis is accepted. However, the probability of a Type II error depends on the specific alternative hypothesis. For a particular H_A this probability is often represented by β. The probability of rejecting a null hypothesis when it is false is called the power of a test. The power is therefore a probability with the value $1 - \beta$, and clearly also depends on H_A and is related to a Type II error. The position is summarised in this table.

	Decision: accept H_0	Decision: reject H_0
H_0 true	Correct decision. The confidence level is 1- α.	Type I error, probability α.
H_A true	Type II error, probability β.	Correct decision, probability 1- β, the power of the test

Calculating the power initially requires specifying the effect size that is required to be detected. (This will be demonstrated in the following example.) The greater the effect size, the greater the power. Power can also be increased by improving the precision of the measurements and by increasing the number of replicated results. Increasing the number of

---

[1] Ethical considerations are involved as well as money. Experiments involving people or animals should be as small as consistent with a useful outcome.

replications is the most commonly used method for increasing statistical power. Although there are no formal standards for power, a value of more than 0.80 is sometimes regarded as satisfactory.

Example

Suppose that we wish to see whether the concentration of an analyte in a test material is significantly different from 20 ppm at the 95% confidence level, so that $\alpha = 0.05$. We consider making four measurements of concentration on the material by using a method with a known standard deviation of 2 ppm. We wish to know whether this experiment would be likely to provide the information that we want. The null hypothesis H_0 for the test is represented by the upper graph in Fig. 7.11.1, which shows the t-distribution with three $(n-1)$ degrees of freedom, with the critical regions for 95% confidence shaded black. The upper limit U of the confidence region falls at $\mu + t\sigma/\sqrt{n} = 23.18$ for data drawn originally from a normal distribution. An observed mean above this limit would occur with a probability of 0.025 for this two-tailed test and would be regarded (falsely) as significant under H_0. Note that we are doing this calculation *before* making any measurements, and have to rely on previous experience of the precision of the analytical method.

Now suppose that, in the context of the application, a deviation of less than 4 ppm from the H_0 value of 20 could be regarded as inconsequential or unimportant. We can then focus on the specific alternative hypothesis $H_A : \mu = 24$ as the upper limit of this acceptable range. This is represented

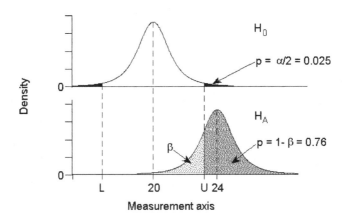

Fig. 7.11.1.   Representation of a null hypothesis (upper graph, $H_0 : \mu = 20$ with lower (L) and upper (U) 95% confidence limits) and an alternative hypothesis (lower graph, $H_A = 24$) showing the power $(1 - \beta)$ of the test.

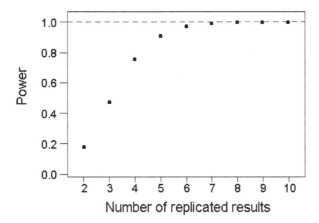

Fig. 7.11.2. Estimated power of the significance test as a function of the number of repeat measurements.

by the lower graph in Fig. 7.11.1. If H_A were true, we would make a Type II error with a probability of $\beta = 0.24$. (This is shown as the area below U, shaded light grey in the lower graph, and calculated from the t_3 distribution). The test would therefore have a power of $(1 - \beta) = 0.76$ (area shaded dark grey in the graph). A material containing 24 ppm would be detected as significantly different in only 76% of experiments as originally proposed, which would be unsatisfactory in many applications.

If we needed a more powerful test we could increase the number of repeat measurements. The relationship between power and number of measurements in this experiment is shown in Fig. 7.11.2. It is clear that a sample of six repeat measurements would nearly always provide the information we wanted (with a power of 0.97), while an experiment with less than five measurements would not. It is a good idea to err on the safe side, as the power of the actual experiment may not be as good as predicted. (We could also increase the power by using a more precise analytical method, if one were available.)

Notes and further reading

- *Power can be estimated for all of the usual tests of significance, including those in analysis of variance and regression.*
- *The logic of estimating power is intricate but statistical software packages usually provide power calculations for the most common tests.*
- *'Significance, importance and power'. (March 2009). AMC Technical Briefs No. 38. Free download via www.rsc.org/amc.*

PART 2

Data Quality in Analytical Measurement

Chapter 8

Quality in Chemical Measurement

This chapter reviews the topic of 'quality' in analytical measurement as a basis for the following discussion of the statistical methods involved. Quality concepts and practices are summarised under three main headings, fitness for purpose, method validation and quality control. However, the overarching idea is the uncertainty attached to an analytical result, what it means, why it is important and how to estimate and use it.

8.1 Quality — An Overview

Key point
— The principles and practices relating to the quality of analytical data can be systematised under three related headings: fitness for purpose, method validation and quality control.

At first sight the number of concepts and practices applied to quality in analytical chemistry is dauntingly large and, moreover, apparently not connected adequately into a coherent whole by overarching principles. However, a simple scheme that provides an inclusive overview can be formulated in terms of just three basic ideas, which should be applied in the following order.

- *Fitness for purpose*: what uncertainty in the analytical result is acceptable to, and best suited for, the needs of the customer?

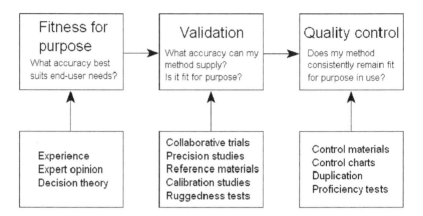

Fig. 8.1.1. A schematic view of the three principal aspects of quality in analytical chemistry and the contributory practices that relate to them.

- *Method validation*: can the method under consideration for the analytical task produce a suitably low uncertainty when executed in a particular environment, in other words, is it apparently fit for purpose?
- *Quality control*: have the environmental factors that determine uncertainty changed since the validation demonstrated that fitness for purpose was achievable? (i.e., did the method work well day after day?)

The logical sequence and the practices that contribute to each are shown in Fig. 8.1.1. Each of these aspects of quality will be considered in turn, but first we have to consider briefly the idea of the uncertainty of a result and how it can be estimated.

8.2   Uncertainty

Key point

— The meanings of the following terms are discussed: *uncertainty, standard uncertainty, expanded uncertainty, coverage factor, measurand* and *traceability*.

The purpose of analysis is to reduce uncertainty about the chemical composition of the test material. Before any analysis is undertaken, we might be in a state of complete uncertainty about what a material is, but that would be unusual. We are far more likely to have some indication of what it might contain *a priori*. For example, a sample of (dried) cabbage would nearly always have a copper content between 1 and 20 ppm. After analysis this uncertainty would be far smaller: we might be confident that the concentration fell, say, in the interval between 10 and 12 ppm. But, however careful the analyst, there would always be some uncertainty remaining in the analytical result. How do we quantify this uncertainty?

First, we need to know exactly what it is that we are estimating: here are the current internationally-recognised definitions.

- *Measurement uncertainty*: non-negative parameter characterising the dispersion of the quantity values being attributed to a measurand, based on the information used.
- *Measurand*: quantity intended to be measured.
- *Quantity*: property of a phenomenon, body or substance where the property has a magnitude that can be expressed by a number and a reference.
- *Standard uncertainty*: measurement uncertainty expressed as a standard deviation.
- *Expanded uncertainty*: product of a combined standard measurement uncertainty and a factor larger than the number one.
- *Metrological traceability*: property of a measurement result whereby the result can be related to a reference through a documented unbroken chain of calibrations, each contributing to the measurement uncertainty.

These formal definitions are not uniformly intelligible and do not immediately convey their meaning. Their application to chemical measurement needs elucidation. *Expanded uncertainty* (U) defines a concentration interval around the result of the measurement within which we expect the true value to lie with a reasonably high probability, usually 95%. The *standard uncertainty* (u) is the basic value that is used to calculate U as $U = ku$, where k is the *coverage factor*, usually between two and three. Standard uncertainty is treated and used in the same way as standard deviation. *Traceability* describes the relationship between the result and the units of the SI (*Le Système International d'Unités*). The *measurand* is a quantity that is being measured (e.g., mass, length, time, concentration), not a chemical substance (that is the *analyte*), nor the numerical outcome of a measurement (that is the *result*).

It is noteworthy that uncertainty is the property of a measurement *result*, while *bias* and *precision* are properties of measurement methods.

Broadly speaking, we should try to use all of these words correctly so as to reduce misunderstanding, especially when our words may be translated into another language. It is especially important not to confuse uncertainty with error.

Further reading

- *Evaluation of measurement data — guide to the expression of uncertainty in measurement (GUM). (2008). Document produced by Working Group 1 of the Joint Committee for Guides in Metrology. This document can be downloaded gratis from the BIPM website www.bipm.org/utils/common/documents.*
- *International vocabulary of metrology — basic and general concepts and associated terms (VIM). (2008). Document produced by Working Group 2 of the Joint Committee for Guides in Metrology. This document can be downloaded gratis from the BIPM website www.bipm.org/utils/common/documents.*
- *Quantifying uncertainty in analytical measurement. (2000). EURACHEM/CITAC Guide CG 4 Second edition. This document can be downloaded gratis via www.eurachem.org/guides.*

8.3 Why Uncertainty is Important

Key points
— Analysis is conducted to inform decisions.
— Logically we cannot make a valid decision without knowing the uncertainty associated with the result.

The result of an analytical measurement is incomplete without a statement (or at least an implicit knowledge) of its uncertainty. This is because we cannot make a valid decision based on the result alone, and nearly all analysis is conducted to inform a decision. Typical decisions based on analysis mostly come in one of the following forms. They all require a knowledge of uncertainty for a rational outcome.

- Does this batch of material contain less than the maximum allowed concentration of an impurity?

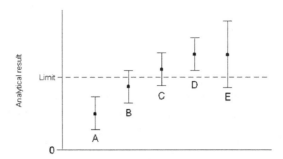

Fig. 8.3.1. Possible results of an analysis (solid circles) and expanded uncertainties (vertical bars) in relation to a legal or contractual upper limit for the concentration of an impurity.

- Does this batch of material contain at least the minimum required concentration of a named ingredient?
- How much is this batch of material worth?

Figure 8.3.1 shows a variety of instances affecting decisions about externally imposed limits. The error bars can be taken as expanded uncertainties, effectively intervals containing the true value of the concentration of the analyte with 95% confidence.

Result A clearly indicates a material that is below the limit, as even the highest extremity of the uncertainty interval is below the limit. Result B is below the limit but the upper end of the uncertainty is above the limit, so we are not sure that the true value is below. Result C is above the limit but the lower end of the uncertainty is below the limit, so we are not sure that the true value is above. It is interesting to compare the equal results D and E. Both results are above the limit but, while D is clearly above the limit, E is not so because the greater uncertainty interval extends below the limit. Organisations affected by such decisions have to agree in advance how to act upon results B, C and E.

Notes and further reading

- *Accreditation agencies require estimates of uncertainty before a method can be accepted as validated.*
- *Use of uncertainty information in compliance assessment. (2007). EURACHEM/CITAC Guide. This document can be downloaded gratis via www.eurachem.org/guides.*
- *The main normative documents on uncertainty are listed in §8.2.*

8.4   Estimating Uncertainty by Modelling the Analytical System

Key points

— Uncertainty can be estimated by the metrological ('bottom-up') method by creating a complete model of the measurement procedure and combining the fundamental uncertainties of the ultimate operations.

— In chemical analysis, modelling often gives a value that is too small because there are nearly always unidentifiable sources of error.

Chemical measurement usually involves a complex multistage procedure. Each stage of the procedure is potentially subject to variation in execution, and therefore makes its own contribution to the uncertainty of the result. If the procedure can be completely characterised as a statistical model, the uncertainties related to each separate stage can be estimated and combined to give the uncertainty of the result. These contributions are best seen as hierarchical. So the determination of copper in a sample of cabbage could be modelled as Fig. 8.4.1 as a first stage.

Each of the three first-level contributions can be further broken down, as exemplified in Fig. 8.4.2 for one of them.

Even more detail can be built into the model (Fig. 8.4.3): for example, weighing introduces uncertainty in the calibration of the weights (or balance) and in correction for buoyancy, absorption of moisture from the atmosphere, and so on. Ultimately the calibration of the weights can be traced back to the SI unit of mass, the kilogramme. Such ultimate

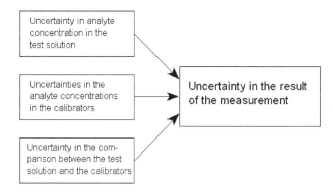

Fig. 8.4.1.   First level of factors that contribute to uncertainty in an analytical result.

Fig. 8.4.2.   Second level of factors that contribute to uncertainty in the analyte concentration in the test solution.

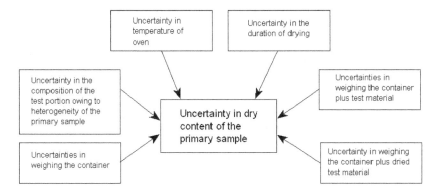

Fig. 8.4.3.   Third level of factors that contribute to uncertainty in the dry content of the primary sample.

traceability is seldom if ever a practical issue for analytical chemistry. Transfer of the SI unit to the analyst's bench gives rise to a negligible proportion of the analytical uncertainty in nearly all instances.

When the measurement procedure is completely broken down in this fashion the influences of the individual uncertainties are estimated in various ways and combined in the manner prescribed by error propagation theory (§8.5) to give the uncertainty of the result. This is clearly a lengthy operation, although simplified by the fact that, because of the way uncertainties are propagated, small uncertainties provide a negligible contribution to the outcome.

This approach to uncertainty is called the 'cause and effect' method or, informally, the 'bottom-up' method. A benefit of the method is that spreadsheets are available that can carry out the calculations once the model is defined. The effect of a change in one of the factors contributing to uncertainty can be rapidly seen in the combined uncertainty. A drawback is that it is difficult to detect structural mistakes and omissions in the model itself. In chemical measurement this defect often results in estimates of uncertainty that are too low. We know this to be the case by studying the results of interlaboratory comparisons. If individual uncertainty estimates were correct, they would account for all of the observed interlaboratory variation. In cases where this has been checked, it is nearly always found that interlaboratory variation is greater than expected on the basis of individual uncertainties. This demonstrates that there are typically unknown causes of uncertainty in chemical analysis. An unknown cause cannot be included in a 'cause and effect' model.

Notes and further reading

- *This method of estimating uncertainty is covered in detail in the 'GUM' and the Eurachem Guide (see §8.2).*
- *Ellison, S.L.R. and Mathieson, K. (2008). Performance of Uncertainty Evaluation Strategies in a Food Proficiency Scheme, Accred. Qual. Assur., **13**, pp. 231–238.*

8.5 The Propagation of Uncertainty

Key points

— Simple mathematical rules are available for combining intermediate uncertainties contributing to a final result.
— Broadly speaking, the outcome will be dominated by the major uncertainties.

The propagation of uncertainty through calculations from intermediate measurement results (A, B, C, etc.) to the final result (x) is handled in the same way as general error propagation. The mathematical rules for combining independent features are as follows. The first three are of greatest importance to analytical chemists.

1. If $x = A \pm B \pm C \pm \cdots$, with respective standard uncertainties $u_A, u_B, u_c, \cdots$ (i.e., for addition or subtraction), the standard uncertainty on x is given by $u_x = \sqrt{u_A^2 + u_B^2 + u_C^2 + \cdots}$.
2. If the results are multiplied or divided, it is the relative uncertainties that are combined. So if $x = \dfrac{A \times B \times \cdots}{C \times D \times \cdots}$, then

$$\frac{u_x}{x} = \sqrt{\left(\frac{u_A}{A}\right)^2 + \left(\frac{u_B}{B}\right)^2 + \left(\frac{u_C}{C}\right)^2 + \left(\frac{u_D}{D}\right)^2 + \cdots}.$$

As a special case of this, if $x = A^k$, k being a constant, then $\dfrac{u_x}{x} = \dfrac{\sqrt{k}u_A}{A}$.

3. If $x = kA$, k being a constant, then $u_x = ku_A$.

4. Generally, if $x = f(A)$, then $u_x = \left| u_A \dfrac{dx}{dA} \right|$.

These rules are used sequentially when applied to complex equations.

As an example we consider the standardisation of hydrochloric acid by titration of a weighed quantity of pure sodium carbonate. The primary measurements are as follows.

- First weighing of sodium carbonate (w_1) : 15.2086 g.
- Second weighing (w_2) : 15.0501 g.
- Uncertainty of a single weighing (u_w) : 0.0002 g.
- Initial burette reading (v_1) : 0.53 ml.
- Final burette reading (v_2) : 38.83 ml.
- Uncertainty in a single burette reading (u_v) : 0.03 ml.
- Uncertainty in end-point recognition (u_e) : 0.04 ml.
- Relative molecular mass (R) of Na_2CO_3 : 106.00.
- Uncertainty of the relative molecular mass (u_R) : 0.01.

The concentration of the acid $(mol\,l^{-1})$ is given by $\dfrac{2000(w_1 - w_2)}{R(v_2 - v_1)} =$ 0.07808.

By Rule 1 the uncertainty on the weight (W) of Na_2CO_3 is given by $u_W = \sqrt{2u_w^2} = \sqrt{2 \times 0.0002^2} = 0.00028$. (Note that there are two weighings to obtain this weight.)

By Rule 1, the uncertainty on the volume (V) of acid is given by $u_V = \sqrt{2u_v^2 + u_e^2} = \sqrt{2 \times 0.03^2 + 0.04^2} = 0.0583$. (Note that there are two volume readings *and* the uncertainty on the end-point recognition to account for.)

As the remaining operations are multiplication and division, we use Rule 2 to complete the calculation.

$$\frac{u_M}{M} = \sqrt{\left(\frac{u_W}{W}\right)^2 + \left(\frac{u_V}{V}\right)^2 + \left(\frac{u_R}{R}\right)^2}$$

$$= \sqrt{\left(\frac{0.00028}{0.1585}\right)^2 + \left(\frac{0.0583}{38.3}\right)^2 + \left(\frac{0.01}{106}\right)^2}$$

$$= \sqrt{(3.12 \times 10^{-6} + 2.32 \times 10^{-6} + 9 \times 10^{-9})}$$
$$= 0.00233.$$

$$u_M = 0.07808 \times 0.00233 = 0.00018 \,\text{mol}\,\text{l}^{-1}$$

(Note that the uncertainty term for the relative molecular mass is negligible.)

Thus, the acid concentration is $0.07808 \,\text{mol}\,\text{l}^{-1}$ with a standard uncertainty of 0.00018.

Notes and further reading

- *If the features are not independent, the covariances have to be taken into account. Details can be found in GUM (§8.2).*
- *It is clear from the way that uncertainties combine that a single contribution that is less than a quarter of the dominant contribution will make negligible contribution to the combined uncertainty. We see that $\sqrt{u^2 + (0.25u)^2} \approx 1.03u \approx u$. Estimates of uncertainty will hardly ever be as accurate as two significant figures, so the difference between $1.03u$ and u is negligible. This will often simplify the calculations in 'bottom-up' estimation.*

8.6   Estimating Uncertainty by Replication

> **Key point**
> — The reproducibility standard deviation is easily obtained and is often a serviceable ('top-down') estimate of uncertainty.

An alternative way of looking at uncertainty is to attempt to replicate the whole analytical process and calculate the uncertainty as the standard deviation. This is called the 'top-down' approach. However, the standard deviation will not be a good estimate of the uncertainty unless two conditions are fulfilled.

- First, there must be no perceptible bias in the procedure, that is, the difference between the expectation of the result and the true value must be negligible in relation to the standard deviation. This condition is usually (but not always) fulfilled in analytical chemistry.
- Second, the replication has to explore all of the possible variations in the execution of the method (or at least all of the variations of important

magnitude). This latter condition cannot usually be met by replication under repeatability conditions (i.e., within-laboratory, see §9.2), because variations in execution of the procedure would be laboratory-specific to a substantial extent. This fact is clearly seen in the results of collaborative trials (§9.7, 9.8), where the reproducibility standard deviation σ_R is on average about twice the repeatability standard deviation.

Experiments have shown that the between-laboratory standard deviation σ_R is often a good estimate of uncertainty, better than many laboratories estimate from within-laboratory validation exercises. Indeed, estimates of standard uncertainty less than σ_R should be considered suspect unless the laboratory concerned can provide evidence of unusually careful procedures or special methods, such as might be found in national reference laboratories. However, σ_R may well be an *under*-estimate of uncertainty in other laboratories. Therefore, laboratories claiming an uncertainty equal to σ_R should attempt to justify that, for instance, by reference to the results from proficiency tests (§11.1).

In practice, laboratories seldom use pure bottom-up or top-down methods for estimating uncertainty, but more often a combination of the two.

Further reading

- *Ellison, S.L.R. and Mathieson, K. (2008). Performance of Uncertainty Evaluation Strategies in a Food Proficiency Scheme, Accred. Qual. Assur., 13, pp. 231–238.*

8.7 Traceability

Key points

— The traceability of a result shows that its units are properly related to the corresponding SI units with an appropriate uncertainty.

— The uncertainty involved in delivering the SI unit to the analyst's bench is usually negligible.

— Other sources of uncertainty tend to be dominant in chemical measurement.

Traceability for a result shows how any unit in which the result is expressed is compared with the parent SI unit. To express a result in (say) moles per

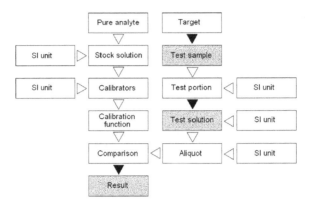

Fig. 8.7.1. Steps showing how an analytical result is traceable to its ultimate origins. The black arrows leading to shaded boxes indicate actions where relatively large uncertainties may be incurred.

litre, the analyst has to know what a mole is, and what a litre is, and show that the measurement comprises a complete chain of comparisons from the definition of the unit to the result. For analytical chemists the necessary connection with SI units is easy. The relative uncertainty in the journey from the SI unit to the analyst's bench is small and nearly always negligible in comparison with the relative uncertainty in the final result. This is because the dominant sources of error in analysis come from elsewhere. The most obvious are shown in the schematic of an analytical measurement in Fig. 8.7.1.

- The sampling error is introduced by taking the sample from the 'target', the bulk of material of which we need to know the composition (§12.1). Sampling errors can be relatively large and sometimes exceed those incurred during the whole of the remaining analytical procedure.
- Preparing the test solution from the test portion often involves chemical manipulations, such as dissolution and separation, in which the recovery of the analyte is incomplete. Correcting for incomplete recovery (or failing to correct) can introduce a relatively large uncertainty in some instances.
- Comparing concentrations via the calibration function and the analytical signal from the test solution is subject to extra error (i.e., beyond that described in §8.4) if there is an unrecognised matrix mismatch with the calibrators.

In addition to these 'non-SI' comparisons, we have to recognise that in many instances the result is calculated from *ratios* of measurements made in the same unit, and in forming such a ratio the link with the SI unit is annulled. For instance, if we are expressing our result as a mass fraction, the most commonly required outcome of chemical measurement, we would get the same numerical result (within limits determined by random variation) if we used a different base unit for mass, the Imperial pound for instance. Such a result can hardly be said to be traceable to the kilogramme even when, as usual, SI weights are used throughout the procedure.

Further reading

• *Traceability in chemical measurement. (2003). Eurachem/CITAC Guide. This document can be downloaded gratis via www.eurachem.org/guides.*

8.8   Fitness for Purpose

Key points
— A fit-for-purpose result is optimal for the end-user.
— The uncertainty of a result is inversely related to its cost.
— The probability of incorrect decision based on a result is directly related to uncertainty.

An analytical result can be said to be fit for purpose when its uncertainty is optimal. The optimality stems from the balance between the cost of conducting the analysis and the cost and probability of making an incorrect decision based on the result. Such a balance is usually determined by expert opinion on the basis of experience of the application. In favourable instances (i.e., where sufficient information about costs etc. is available) it may be possible to calculate the optimum directly. Even when this cannot be done, decision theory provides a useful conceptual framework for a more transparent estimate.

The cost of an analytical result (including sampling where appropriate) is related to the uncertainty required. Smaller uncertainty costs more money. Indeed, a useful rule of thumb is that of an inverse square relationship, that is, a procedure that reduces uncertainty by one half costs four times the amount. The probability of an incorrect decision also depends on the uncertainty, the greater the uncertainty the greater the probability of an incorrect decision and therefore of a financial penalty stemming from the decision. Such penalties might be the result of mistakenly condemning

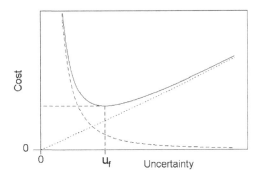

Fig. 8.8.1.   Costs versus uncertainty, showing costs of measurement (dashed line), costs of incorrect decisions (dotted line) and total costs (solid line). The uncertainty u_f at minimal cost is regarded as defining fitness for purpose.

a batch of satisfactory material, or incorrectly shipping a batch of material that was out of specification.

The sum of the two costs (one a decreasing function of uncertainty and the other increasing) must necessarily have a minimum value, and this point provides a rational definition for a fit-for-purpose uncertainty (Fig. 8.8.1). Of course, an uncertainty smaller than that demanded for fitness for purpose would give an even smaller proportion of mistaken decisions, but would be unnecessarily expensive.

Notes and further reading

- *Fearn, T., Fisher, S., Thompson, M. et al. (2002). A Decision Theory Approach to Fitness for Purpose in Analytical Measurement, Analyst,* **127***, pp. 818–824.*
- *'Optimising your uncertainty — a case study'. (July 2008). AMC Technical Briefs No. 32. Free download via www.rsc.org/amc.*

Chapter 9

Statistical Methods Involved
in Validation

The statistical methods involved in validation are straightforward, but applying them effectively is more involved than generally appreciated. Regression is the natural method to apply to analytical calibration, but a good design is needed for an informative outcome. Particular attention is needed in selecting the appropriate conditions under which precision is estimated for various purposes. Moreover, setting proper control limits for control charts involves a moderately long series of measurements and review at suitable intervals.

9.1 Precision of Analytical Methods

Precision is the smallness of variation in the results of replicated measurements. It is characterised in terms of standard deviation, so high precision is equivalent to low standard deviation. Analytical results for different purposes vary in precision: a relative standard deviation (RSD) of 0.1% is 'high precision' in analysis and is seldom attained except for special purposes, such as finding the commercial value of materials containing gold. Most analytical results have repeatability RSDs of about 1–5% except in the measurements of very low concentrations, where even higher levels prevail. RSDs higher than about 30% are problematic, because the expanded uncertainty is of the same magnitude as or greater than the result.

There is little technical difficulty about estimating a standard deviation — the measurement has to be replicated under appropriate conditions and the standard deviation of the results calculated, in the absence of complications, as $s = \sqrt{\sum_i (x_i - \bar{x})/(n-1)}$. The analyst should, however, initially examine the dataset to ensure the absence of features that could create a false impression, such as outlying data points (see §7.2 ff) or trends.

A simple plot of the results in the order measured is usually enough to show such problems. If appropriate, the influence of outliers can be accommodated by the use of a robust estimate (§7.6).

The effect of trends, even complicated ones, can also be reduced by various methods. Perhaps the simplest is to consider the signed differences between successive results, that is, the sequence $d_1 = x_1 - x_2$, $d_2 = x_2 - x_3, \ldots, d_{n-1} = x_{n-1} - x_n$. The standard deviation of the signed differences divided by $\sqrt{2}$ is equal to that of the original data de-trended. In the following example (Fig. 9.1.1), the raw results (upper series) show a strong trend and have a standard deviation of 3.02. The signed differences show no trend and have a standard deviation of 1.59. The de-trended data would therefore have an estimated standard deviation of $1.59/\sqrt{2} = 1.12$.

Another important feature of estimates of standard deviation is that they are very variable with small to moderate numbers of observations. The standard error $se(s)$ of an estimate s based on n normally-distributed results is given by $se(s) = \sigma/\sqrt{2n}$. With the usual ten replicated analytical results, we could expect to see 95% confidence intervals of around ±40% around the estimated standard deviation. (Note: the confidence limits will not be symmetrically disposed around s for small n.) Two estimates of the same standard deviation based on separate sets of ten results will differ by more than 30% half of the time and by more than 77% in one in ten experiments. Because of this it is rarely worth quoting standard deviations or uncertainties to more than two significant figures, and there is no point in discriminating between minor gradations in precision.

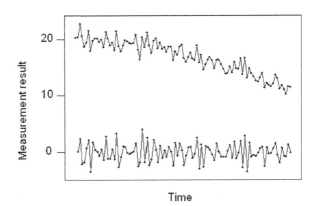

Fig. 9.1.1. Replicated results as a time series (upper plot) and as successive signed differences (lower plot).

Notes and further reading

- *The dataset is available from the file named* **Drift**.
- *There is often a practical problem in estimating the standard deviation when the concentration of the analyte is close to zero. A proportion of the results (sometimes a substantial proportion) will fall below zero unless they are censored. A true concentration cannot fall below zero, but the result of a measurement can. Recording such results as zero, or repeating the measurement until a non-negative result is obtained, will bias the estimates of both the mean (upwards) and the standard deviation (downwards). There are statistical techniques (maximum likelihood estimation) for handling this situation, but they are beyond the scope of this book.*
- *'Measurement uncertainty and confidence intervals near natural limits'. (2008). AMC Technical Briefs No. 26A. Free download from www.rsc.org/amc.*
- *Analytical Methods Committee. (2008). Measurement Uncertainty Evaluation for a Non-negative Measurand: an Alternative to Limit of Detection, Accred. Qual. Assur., 13, pp. 29–32.*

9.2 Experimental Conditions for Observing Precision

The precision of the results of a method depends on the conditions under which the measurement is replicated. There are many conceivable conditions and, regrettably, the terminology used is often confusing (Table 9.2.1). However, only a few of these conditions have a practical bearing on quality practice. The key point here is to identify conditions that are genuinely useful to analytical chemists. Of these, reproducibility precision is an important consideration in method validation, as it is a prominent feature of estimating uncertainty. Run-to-run precision and repeatability precision are mainly relevant to internal quality control.

Further reading

- *International vocabulary of metrology — basic and general concepts and associated terms (VIM). (2008). Document produced by Working Group 2 of the Joint Committee for Guides in Metrology. This document can be downloaded gratis from the BIPM website www.bipm.org/utils/common/ documents.*

Table 9.2.1. Conditions for assessing precision and the utility of the resulting estimates. (Names defined in normative documents are in boldface type.)

Names	Conditions of replication	Comments
Instrumental	Replication as quickly as possible, with no change of test solution nor adjustment of instrument.	Not very useful but often seen in research papers and brochures. Does not include variation originating from chemical manipulations.
Repeatability	Replication on separate test portions of the same material, with the same instrument and reagents, in the same laboratory, by the same analyst, in a 'short' period of time.	The 'short period of time' is the length of an analytical 'run', or period in which we assume that the factors affecting error have not changed. Limited applicability, but used in restricted types of quality control.
Run-to-run: **Intermediate**: Within-laboratory reproducibility	Replication in separate runs. Same method and laboratory, but may be different analysts, instruments and batches of reagent.	This type of precision is addressed in internal quality control.
Reproducibility (1)	Replication by the same detailed method in different laboratories.	This is the estimate provided by the collaborative (interlaboratory) trial.
Reproducibility (2)	Replication by the same nominal method but with variation in details in different laboratories.	Estimate can often be obtained from the results of a single round of a proficiency test. The SD may be greater or smaller than that of Reproducibility (1).
Reproducibility (3)	Replication by various methods in different laboratories.	Estimate can often be obtained from the results of a single round of a proficiency test. The SD is usually greater than that of Reproducibility (1).

9.3 External Calibration

Regression is the natural way to explore the behaviour of an analytical calibration and its likely effect on the uncertainty of the result (see Chapters 5 and 6). Topics that can be studied by regression are as follows.

- Does the calibration depart significantly from linearity? If it does, is the deviation of a magnitude that will affect the uncertainty of the results of a complete measurement, that is, including the chemical treatment of the test portion? This is best addressed by measuring responses in duplicate from calibrators equally spaced over the range and conducting a test for significant lack of fit. Simply inspecting a plot of the residuals for a curved pattern is also a powerful way of detecting lack of fit. However, the correlation coefficient commonly used as a test for linearity is ambiguous in this context and potentially misleading (§5.9). An important aspect of unacknowledged calibration curvature is that its effects on estimated concentrations will be relatively very large at concentrations near zero. If low concentrations are to be measured, lack of fit cannot be tolerated.
- If there is no lack of fit to the selected calibration function, is there a significant intercept? In other words, in a calibration function such as $r = \hat{\beta}_0 + \hat{\beta}_1 c$, are there grounds to reject $\beta_0 = 0$? If there are not, the analyst can use the hypothesis that the calibration passes through the origin.
- Do the residuals display signs of heteroscedasticity? This would be a quite common occurrence in analytical calibration and suggests that the use of weighted regression would give better results than simple regression.
- What are the confidence limits on an unknown concentration found by using the calibration function (also called 'evaluation limits', or 'fiducial limits')? It is salutary to calculate these limits as they are considerably larger than expected by intuition and may make a substantial contribution to the combined uncertainty of the result.
- What is the detection limit of the calibration? When the evaluation confidence interval includes zero concentration, the evaluated concentration is not significantly greater than zero, so we are unsure whether it is present at all.

But how does all this information about calibration relate to the combined uncertainty of a result? The shape of the calibration function is of great importance. If a rectilinear function can be assumed, calculations become simpler, errors of interpolation become smaller and the method of standard additions (§9.5) becomes available. Heteroscedastic residuals are a quite common occurrence in chemical measurement, and their presence suggests that the use of weighted regression would give better results than simple regression. As a rule of thumb, if the calibration is to be restricted to concentrations up to about 200 times the detection limit, simple regression will be good enough. For concentrations outside this range, weighted regression will give more accurate values, especially at low concentrations. Lack of fit if present may be judged negligible, but care must be taken (see §5.10) if we need to measure concentrations in the lower quartile of the range.

We must bear in mind that in analysis there are many sources of uncertainty other than calibration and evaluation, and these will usually overwhelm calibration uncertainty. In such instances, calibration uncertainty *per se* can be ignored or subsumed into uncertainties that are readily estimated by replication.

9.4 Example — A Complete Regression Analysis of a Calibration

Here we examine the calibration of zinc by atomic absorption spectrometry, with duplication of responses, which were measured in the random order listed.

Concentration mg l^{-1}	Absorbance
4	0.237
0	0.004
5	0.291
3	0.177
2	0.124
1	0.061
2	0.124
1	0.064
3	0.182
5	0.300
0	0.009
4	0.241

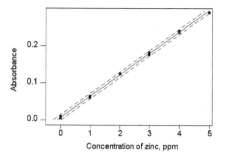

Fig. 9.4.1. Calibration data (points) with linear function estimated by regression (solid line) and 95% evaluation limits (dashed lines).

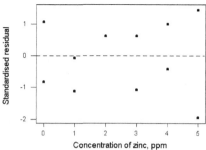

Fig. 9.4.2. Residuals from the regression.

The data show no apparent deviation from the regression line (Fig. 9.4.1). The residuals plotted against concentration show no obvious pattern (Fig. 9.4.2) and the test for lack of fit provides a p-value of 0.867. Thus we can conclude that there is no significant lack of fit and that the regression has provided a satisfactory account of the data.

Source of variation	Degrees of freedom	Sum of squares	Mean square	F	p
Regression	1	0.11774	0.11774	12570.11	0.000
Residual error	10	0.00009	0.00001		
Lack of fit	4	0.00002	0.00000	0.30	0.867
Pure error	6	0.00008	0.00001		
Total	11	0.11783			

Visually the regression line passes close to the origin. As there is no observed lack of fit, the p-value of 0.003 for the intercept shows that we can reject $H_0 : \alpha = 0$ in favour of $H_A : \alpha \neq 0$, that is, the intercept is significantly different from zero. The deviation of 0.006 absorbance would be large for modern instrumentation, but presumably arose from an instrumental drift after the initial setting of the zero point.

Predictor	Coefficient	Standard deviation	t	p
Constant	0.006167	0.001566	3.94	0.003
Concentration	0.0580000	0.0005173	112.12	0.000
	$s_{y|x} = 0.0031$	$R^2 = 99.9\%$		

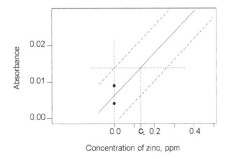

Fig. 9.4.3. Bottom end of the estimated calibration function, showing the detection limit c_L of the calibration.

The 95% confidence interval for a predicted concentration (the evaluation interval) amounts to about 0.3 ppm (mg l^{-1}) over the whole of the calibrated range (Fig. 9.4.1). In effect, that would be the contribution of the calibration procedure to the combined uncertainty. By zooming in to the bottom end of the calibration (Fig. 9.4.3), we see that the calibration detection limit c_L will be about 0.13 mg l^{-1}.

Note

- *An unusual feature of this calibration dataset demonstrates the value of a thorough examination of the residuals. If the residuals are plotted against the order in which the measurements were made (Fig. 9.4.4), we see a systematic effect in that the earlier residuals tend to be negative and the later ones positive. This is a significant effect ($p = 0.004$) and shows that the instrument is still drifting during the calibration. This effect would not have been detectable had the calibrators been analysed in order of increasing concentration.*

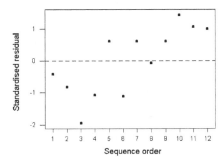

Fig. 9.4.4. Residuals plotted against the order in which the solutions were measured.

9.5   Calibration by Standard Additions

To make a valid comparison between calibrators and test solutions, the intercept and sensitivity of the calibration function must match those we would observe in the matrix of the test solution. A sufficiently-close matrix matching is readily contrived in respect of reagents such as mineral acids added during the chemical treatment of the test portion. In some instances, however, the matrix of the test materials (that is, all of the constituents other than the analyte) is not readily predictable and the matrix varies enough within a class of material to prevent matrix matching being complete. This can lead to an unacceptable addition to the uncertainty of the result if the analytical signal is sensitive to such changes. A number of strategies are available to overcome such interference effects.

Calibration by standard additions is a widely applicable method that overcomes changes in sensitivity (or rotational effects) caused by the matrix (Fig. 9.5.1). The method, however, does not overcome changes to the baseline of the signal, which the analyst has to deal with separately. Standard additions, therefore, has to be applied to the net signal: it requires *any* baseline signal to be subtracted from the gross signal before the method can be applied.

The conventional paradigm of the method, as presented in most textbooks, requires the addition to a test solution of several different exactly-known amounts of the analyte. This has to be done in such a way that the overall concentration of the test material remains the same in all of them, so that the matrix is identical in each solution. The analytical response (corrected for baseline shift) is measured for each solution, and the line fitted to the points is extrapolated down to zero

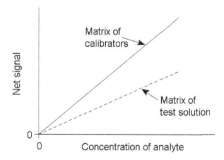

Fig. 9.5.1.   The possible effect of matrix mismatch on the comparison between calibrators and test solution.

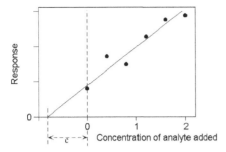
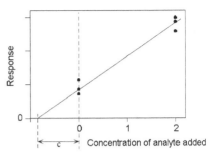

Fig. 9.5.2. The method of standard additions (conventional paradigm), with five different added concentrations of analyte.

Fig. 9.5.3. Standard additions with the spike added at only one level (triplicate measurements of response).

net response (Fig. 9.5.2). The negative reading on the concentration axis is the concentration estimate. The extrapolation could be done graphically, but the application of regression to estimating the original concentration is obvious. The resulting function $y = a + bx$ with the response (y) set at zero gives $x = -a/b$.

Standard additions is a valid method when the calibration is known to be linear. Non-linear extrapolation is unwise (§6.3). The standard paradigm of the method, with several different levels of added analyte (Fig. 9.5.2), is featured in most texts because it is thought (wrongly) to allow the analyst to check that the calibration is linear. However, standard additions should not be attempted unless non-linearity is known to be undetectable by previous experimentation during validation. In any event, testing for non-linearity is unlikely to be fruitful without an inordinate number of measurements to obtain just one result.

As we can assume linearity, a simpler experimental design is preferable, with only one level of added analyte, perhaps with replicated measurement (Fig. 9.5.3). The added level should be the highest possible concentration consistent with a linear calibration function. This design not only cuts down the analyst's workload, but also improves the precision of the final result for the same number of measurements. Moreover, regression is not needed in the calculations. A line passing through the means of the responses at both concentrations is identical with a regression line, either simple or weighted. Standard additions is sometimes regarded as problematic because the extrapolation step was thought to introduce extra imprecision in comparison with eternal calibration. A careful study, with realistic models of uncertainty, has shown that this fear is unfounded.

Further reading

- *'Standard additions — myth and reality'. (2009). AMC Technical Briefs No. 37. Free download via www.rsc.org/amc.*
- *Ellison, S.L.R. and Thompson, M. (2008). Standard Additions: Myth and Reality. Analyst, **133**, pp. 992–997.*

9.6 Detection Limits

Key points

— There are numerous possible definitions of detection limit. The simplest is: the concentration c_L of analyte that corresponds with a signal of $R_0 + 3\sigma$ where R_0 and σ describe the variation in the analytical signal when the actual concentration of analyte is zero.
— Modern more complicated definitions provide very similar estimates.
— Detection limits provide only a rough guide to method performance and should not be taken too seriously.
— Detection limits cited in the literature and in instrument manufacturers' brochures may be misleadingly low.

A detection limit c_L is the smallest concentration of analyte that can be reliably detected by the analytical system. Detection limits are usually given undue attention in relation to their usefulness. The main use of a detection limit is to warn the analyst of a concentration level better to avoid if at all possible. However, in the determination of undesirable contaminants — a very common activity — analysts often have to work near detection limits.

Detection limits are encountered when the expanded uncertainty of measurement is roughly comparable with the concentration of the analyte. But there are complications that affect this basically simple idea.

- There are several different ideas about how the detection limit can be defined in statistical terms. All of these ideas are based on the standard deviation of replicated results near zero concentration.
- The magnitude of the detection limit estimate will depend on the conditions of replication of the measurements. Detection limits quoted in descriptions of methods and instrument brochures are often far too small because they are estimated under unrealistic conditions of replication

such as 'instrumental conditions' (§9.2). Real-life detection limits (based on reproducibility statistics) are sometimes as much as 10 times higher than these instrumental detection limits.

- The accuracy of an estimated detection limit will be poor because it is typically based on ten replicated measurements, while the standard error of an estimated standard deviation is given by $\sigma/\sqrt{2n}$. With $n = 10$ measurements the 95% confidence limits on s (and therefore on c_L) will fall at about $\pm 40\%$ of the true value.
- Detection limits give rise to an artifactual dichotomy of the concentrations axis that distorts perception of reality. Many analysts and end-users alike regard a result of $1.1c_L$ as a valid result and $0.9c_L$ as invalid. Modern thought is moving to the position that detection limits are unnecessary — all that the end-user needs is the result and its uncertainty.

The simplest (and therefore the most commendable) definition of detection limit is this: the concentration c_L of analyte that corresponds to an analytical signal of $R_L = R_0 + 3\sigma$ where R_0 (mean) and σ (standard deviation) describe the distribution of the analytical signal when the actual concentration of analyte is zero. We can see the meaning of this by reference to a short calibration graph (Fig. 9.6.1). The normal distribution describes the variation in the analytical signal (response) when the blank solution is repeatedly presented for measurement. A response larger than $R_0 + 3\sigma$ will occur rarely if no analyte is present (about one time in a thousand, as we are dealing with one-tailed probabilities), so if we saw

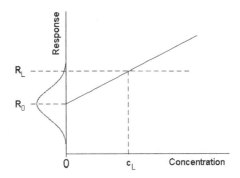

Fig. 9.6.1.   Schematic diagram of a calibration function at low concentrations showing the variation in the response (analytical signal) for zero concentration of analyte, and the detection limit c_L.

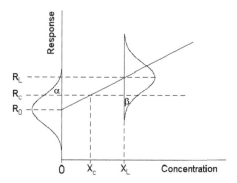

Fig. 9.6.2. Schematic diagram of a calibration function at low concentrations of analyte, showing a more complex definition of detection limit.

a response greater than that we could be confident that some analyte was actually present. We could say that, at that point, the concentration of analyte was significantly greater than zero at a confidence level of about 99.9%.

International standards nowadays prefer a more complex idea (and terminology) of detection limit. For this purpose we consider the previous calibration graph in slightly more detail (Fig. 9.6.2). As before we have the distribution of responses at zero concentration with mean R_0 and standard deviation σ. We now arbitrarily define a 'critical level of response' R_c that defines a probability α in the upper tail of the distribution. This corresponds via the calibration function to a 'critical concentration' x_c. If we looked at the distribution of responses at concentration x_c, half of the responses would be below R_c and therefore, in some sense, 'not detected'. x_c is clearly too low to act as a serviceable detection limit. However, at some higher concentration, where the response was R_L, the area β in the tail of the distribution below a response R_c could be made suitably small to define a sensible detection limit x_L. In practice, we usually make both α and β equal to 0.05, so that we have

$$R_c = R_0 + 1.63\sigma,$$

$$R_L = R_0 + 3.26\sigma.$$

x_L therefore corresponds closely with the previous definition of detection limit c_L.

Further reading

- *Capability of detection — Part 1: Terms and definitions. (1997). 11843-1. International Standards Organisation, Geneva.*
- *'Measurement uncertainty and confidence intervals near natural limits'. (2008). AMC Technical Briefs No. 26A. Free download from www.rsc.org/amc.*
- *Analytical Methods Committee. (2008). Measurement Uncertainty Evaluation for a Non-negative Measurand: an Alternative to Limit of Detection, Accred. Qual. Assur., 13, pp. 29–32.*

9.7   Collaborative Trials — Overview

Key points

— A collaborative trial is an interlaboratory study to characterise the performance of an analytical method.

— The main performance features are precisions of repeatability and reproducibility.

Collaborative trials are interlaboratory studies to characterise the performance of a well-defined analytical method applied to a well-defined type of test material. The performance features characterised are repeatability precision, reproducibility precision and, if certified reference materials are included, trueness. Usually each of these features will vary as a function of concentration of the analyte, so the tests need to be carried out using at least five different materials of the defined type, with concentrations spanning the relevant range. The organising body selects and prepares these materials and distributes them to the participating laboratories, which should be at least eight in number (and preferably considerably more). The participant laboratories analyse each of the materials in duplicate, preferably 'blind' (i.e., without knowing the identity of the duplicates during the analysis). The participant laboratories should be proficient in the type of analytical test involved.

The participants report the results obtained to the organiser, who then carries out the statistical analysis of the results to estimate the various performance indicators. The basic statistical technique in collaborative trials is one-way analysis of variance applied separately to the results from each material (§4.6). However, there are several refinements that are regarded as essential. An important aspect is the initial removal of results identified as outliers. There is an elaborate procedure for doing that described below (§9.8). Alternatively, an approach based on robust analysis of variance has been found to provide very similar outcomes without resort to outlier tests (but see Note below). The justification for rejecting outliers in collaborative trials is that the resulting statistics are regarded as describing the essential properties of the method, not those of the participant laboratories.

When the standard deviations of repeatability and reproducibility are obtained for the various test materials, it is common practice to attempt to find a functional relationship that describes the relationship between precision and concentration of the analyte (§9.9). This relationship can provide a compact description of the capabilities of the method and a means of interpolation to concentrations other than those actually represented in the study.

The results of collaborative trials are often compared with the well-known 'Horwitz function'. This function stems from an empirical observation about the trend of reproducibility relative standard deviation σ_R as a function of concentration c, in the food analysis sector. Perhaps the most useful formulation of the function is $\sigma_R = 0.02c^{0.8495}$, with both variables expressed as a mass fraction. The trend of results from collaborative trials follows this function closely over seven orders of magnitude (between concentrations of about 10^{-8} and 10%) although it must be stressed that there are both random and systematic deviations from the trend. The function is therefore not necessarily a good descriptor of individual methods. It is, however, often used as an independent fitness-for-purpose criterion in method validation and proficiency testing. In that context it is used to *pre*scribe the performance required, not *de*scribe the performance obtained.

Notes and further reading

- *Robust ANOVA must be used as strictly alternative to outlier deletion. The practice of using both methods and then choosing the outcome that is smaller must be avoided.*
- *'The amazing Horwitz function'. (2004). AMC Technical Briefs No. 17. Free download via www.rsc.org/amc.*
- *Precision of test methods Part 1: Guide for the determination of repeatability and reproducibility for a standard test method. (1979). ISO 5275. International Standards Organisation, Geneva.*

9.8 The Collaborative Trial — Outlier Removal

Key points

— Outliers are conventionally removed from collaborative trial datasets before analysis of variance is carried out.
— This is justified because the study is meant to characterise the method rather than the participant laboratories.

The most frequently used method for purging the initial valid data of outliers is defined in the 1995 IUPAC Harmonised Protocol. This procedure consists of the alternating use of the Cochran and Grubbs tests until no further outliers are flagged or until the proportion of dropped laboratories would exceed two-ninths of the original number of laboratories providing valid data (Fig. 9.8.1).

Cochran test (see §4.10)

First apply the Cochran outlier test (one-tail test at $p = 2.5\%$) and remove any laboratory whose critical value exceeds the tabulated value.

Grubbs tests (see §7.4)

Apply the single value Grubbs test (two-tail test at $p = 2.5\%$) and remove any outlying laboratory. If no laboratory is flagged, then apply the pair value tests (two-tail) with two values at the same end and one value at each end, with $p = 2.5\%$ overall. Remove any result flagged by these tests of which the critical value exceeds the tabulated value. Stop removal when the next application of the test would flag as outliers more than two-ninths of the laboratories. (Note: the Grubbs tests are to be applied one material at a time to the set of replicate means from all laboratories, and not to the individual values from replicated designs because the distribution of all the values taken together is multimodal, not Gaussian, i.e., their differences from the overall mean for that material are not independent.)

Final estimation

Recalculate the ANOVA statistics after the laboratories flagged by the preceding procedure have been removed.

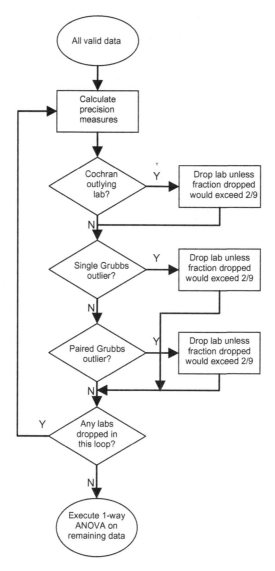

Fig. 9.8.1. Flow chart of procedure for removing outliers from collaborative trial data before ANOVA.

Further reading

- *Horwitz, W. (1995). Protocol for the Design, Conduct and Interpretation of Method Performance Studies, Pure Appl. Chem., **67**, pp. 331–343.*

9.9   Collaborative Trials — Summarising the Results as a Function of Concentration

---

Key points
— Expressing precision data as a function of analyte concentration is a useful way of summarising performance information.
— Finding a good fit is not always straightforward.

It is usually beneficial to summarise the statistics obtained for each material in a collaborative trial by treating the precision as a function of concentration. This provides a compact summary of the study. However, it may not as simple as it first appears. The main problems are as follows.

- There are few data points, and the relative uncertainty in the standard deviation estimate at each point is large. There will be a large relative uncertainty in the estimated functional relationship.

- The data may be markedly heteroscedastic. Some form of weighted procedure should be used (see §6.7).

- Lack of fit may be apparent if there is a significant variation in the matrices of the test materials.

- The intrinsic shape of the true functional relationship is unlikely to be linear and its value must be strictly positive. Standard regression methods may not be applicable, i.e., we cannot use a function that imputes a negative standard deviation.

These features are apparent in Fig. 9.9.1. A number of functional forms are suggested in ISO 5275, but they are all fundamentally flawed. In particular, simple linear regression will be suspect as it will not take account of the heteroscedasticity and tend to give a negative or otherwise misleading intercept.

Some simple methods that automatically take account of the heteroscedasticity may be conditionally appropriate in specific circumstances. When all of the materials have analyte concentrations well above the detection limit of the method, a constant relative standard deviation (RSD) is a reasonable assumption unless the data are clearly at odds with that. In such an instance a suitable expedient might be to calculate the average of the RSDs. Interpolation could then be executed by applying this average value to new concentration values. Alternatively, a similar outcome could be obtained by applying simple regression to the log-transformed data. As

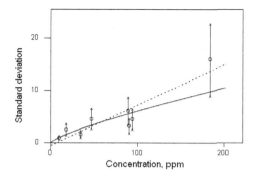

Fig. 9.9.1. Standard deviation vs. concentration (points) for a collaborative trial of a method for determining propyl gallate. The vertical bars show the 95% confidence intervals for the estimate. Two fits are shown, simple linear regression (dotted line) and log-log regression (solid curve). Both functions show lack of fit to some points and indicate inappropriate values near zero concentration.

the model for constant RSD is $\sigma = \beta c$, where σ is the standard deviation at concentration c, the transformed equation is $\log \sigma = \log \beta + \log c$. This function should have a slope of one (unity) if the model is correct, and a t-test on the slope should be able to confirm that. Given a unit slope, the *intercept* would be $\log \hat{\beta}$, and we would calculate $\hat{\beta}$ as its antilog. The value of $\hat{\beta}$ is then the RSD for interpolation, that is, for calculating values of σ from new values of c within the range of the original data and well away from the detection limit.

The method of regressing logarithms will also work if the functional relationship has features similar to the Horwitz function, that is, of the form $\sigma = \beta c^{\gamma}, \gamma \neq 1$ so that $\log \sigma = \log \beta + \gamma \log c$. The regression *coefficient* will then equal the unknown exponent γ, which may or may not differ from the Horwitz exponent of 0.8495 (see §9.7). An example of this method is shown in §6.10.

However, taking logarithms will tend to be misleading if the concentration values are lower than about 100 times the detection limit. The fundamental reason is that the relationship between standard deviation of measurement and concentration of the analyte must have a positive intercept. In other words, the standard deviation at zero concentration *must* be greater than zero, because it is describing a measurement result.

Many analytical systems can be better described by a function of the form $\sigma = \sqrt{\alpha^2 + (\beta c)^2}$, which is technically appropriate as it has an intercept of α and tends towards a constant RSD of β at high concentrations. However, a weighted fitting of this type of function goes beyond elementary statistical considerations.

Further reading

- *Thompson, M., Mathieson, K., Damant, A. P. et al. (2008). A General Model for Interlaboratory Precision Accounts for Statistics from Proficiency Testing in Food Analysis, Accred. Qual. Assur., **13**, pp. 223–230.*

9.10 Comparing Two Analytical Methods by Using Paired Results

> **Key points**
> — Comparison between two analytical methods can be undertaken with paired results, using either a simple *t*-test or a regression-type method.
> — If the comparison is based on results from a single laboratory, the outcome may not be generally applicable.

A common analytical task is the comparison of results from two methods for measuring the concentration of the same analyte in a number n of test materials. Usually one of the methods is recognised as a reference method and the other, perhaps more rapid or convenient, is under trial. This 'paired results' method is a particularly valuable approach because the comparison between methods is based on the behaviour of 'real-life' test materials, not on reference materials or spiked solutions which might behave atypically. The comparison is obviously a statistical matter, but the correct technique for the interpretation of the results depends on the concentration range spanned by the test materials and on the precisions of the two methods. If the concentration range is relatively small it might be possible to assume a single variance for each method, in which case a *t*-test of the differences between corresponding pairs would be suitable (§3.8–§3.10). If the concentration range is wider, it is probably advantageous to make the comparison a function of concentration, as in §5.12. Quite commonly, the reference method will produce results $x_{i=1,...,n}$ that are substantially more precise than the corresponding trial method results $y_{i=1,...,n}$. In such instances regression of y on x (but not x on y) will probably be a suitable statistical technique (§5.12). If the precisions are comparable, however, regression may lead to a misleading interpretation because a basic assumption of regression is violated. In such instances, a more general technique that accommodates variable precision on both variables would be required. This 'functional relationship' fitting will provide an unbiased outcome, but an account of the method is beyond the scope of the present work.

Some care is needed in the design of such experiments, especially in regard to the required generality of the conclusions. Often the reference method and the trial method are quite different in the analytical procedures and the physical principles behind the procedures. In such instances the potential existence of laboratory bias has to be taken into account. Laboratory bias, small or large, is always present in results, as can clearly be seen in results of proficiency tests. Paired results from a single laboratory will not be descriptive of the methods because of the unique biases in that laboratory. In short, the conclusions of the study would apply only to the laboratory concerned. There is, however, an important exception to this limitation: when the two methods differ only in one particular, the effect under study, the laboratory bias (apart from the effect under study) will be the same for both methods and thus cancel out.

If a general (rather than a laboratory-specific) conclusion is required, so that it applies to the methods (and therefore to all laboratories), it is necessary to compare the mean results of a number of laboratories, for both methods and for each material. This requires a large study, comparable in size (and indeed cost) to a double collaborative trial. There is one mitigating circumstance where suitable data can be obtained at no cost, and that is where both methods are well-characterised and well-represented in a large proficiency test. Then it is possible to compare the means of medium-sized datasets (e.g., 20–50 laboratories) and use the respective variances as weights for fitting a functional relationship.

Notes and further reading

- *Requirements for a valid comparison of the results of methods are discussed in Thompson, M., Owen, L., Wilkinson K. et al. (2002). A Comparison of the Kjeldahl and Dumas Methods for the Determination of Protein in Food, Using Data from a Proficiency Testing Scheme, Analyst, **127**, pp. 1666–1668.*

Chapter 10

Internal Quality Control

This chapter is concerned with the statistical aspects of internal quality control, with special emphasis on the correct setting up of control charts. It covers the meaning of statistical control, within-run and run-to-run precision, and the use of multiple-analyte control charts.

10.1 Repeatability Precision and the Analytical Run

> **Key points**
> — Duplication can be used to estimate or control repeatability (within-run) variation.
> — The key statistic is the absolute difference between corresponding duplicated results.
> — It is usually necessary to take account of how precision depends on the analyte concentration.
> — 'Maps' of absolute difference against concentration can utilise control lines for a prescribed repeatability precision.

Repeatability conditions of precision are formally defined as those under which replicate measurements are made on the same test material by the same analyst using the same method, equipment and batch of reagents and within a 'short' period of time (§9.2). The undefined short period can be most usefully interpreted as the length of an analytical 'run', a continuous period, involving anything from one to a large number of measurements,

189

during which factors contributing to the magnitude of errors are deemed
to remain constant. Of course, the conditions are never really constant and
some systematic changes can be expected in a typical run. The effect of
undetected changes of this kind can be handled by conducting the sequence
of analyses in a random order. The errors can then be regarded as part of
the repeatability variation.

In any event, repeatability standard deviation (σ_r) is of only limited
use to analytical chemists as it is usually considerably smaller than the
uncertainty of the measurement. Its main value is in enabling the analyst
to gauge whether results replicated within a run are consistent, with each
other or with some externally-derived criterion. Duplication within a run
thus provides the analyst with a restricted type of quality control, which
is executed by consideration of the difference between the duplicate results
on the test materials. The method has the advantage that the variation
observed is that of materials that are entirely typical, both in composition
and state of comminution. To represent the true variation within the run,
however, the duplicated test portions must be at random positions in the
analytical sequence — if they were adjacent, or simply close in relation to
the length of the run, the variation observed between pairs would tend to
be too small, because they would not account for unremarked systematic
changes.

The test statistic is the difference $d = x_1 - x_2$ between corresponding
pairs of results, and this has a standard deviation $\sigma_d = \sqrt{2}\sigma_r$. The difference d has an expectation (long-term average) of zero only if there is no
consistent trend in the instrumental performance. (For instance, if the sensitivity of the instrument were consistently falling during the run, the first
of a pair of duplicated results would tend to be greater even if the duplicate
test portions were analysed in a random order.) This apparent difficulty can
be overcome simply by considering only the absolute difference $|d|$ between
the corresponding results.

A complication arises when, as often happens, the concentration of
the analyte varies considerably among the test materials comprising the
run. In such an instance we would expect σ_r, and therefore σ_d, to vary
with concentration. If we postulated a functional relationship of the form
$\sigma_r = \sqrt{\alpha^2 + (\beta c)^2}$, we could predict that $\sigma_d = \sqrt{2(\alpha^2 + (\beta c)^2)}$. If we knew
or prescribed values for the parameters α, β, we could calculate σ_d at any
concentration c by inserting $c = (x_1 + x_2)/2$ into the equation. Under

the normal assumption, the scaled differences d/σ_d should behave like a sample from a standard normal distribution, regardless of any varying concentration in the materials. There is no point in plotting these results on a Shewhart control chart because the results are not a temporal sequence. A dotplot is sufficient to show the distribution.

More information, however, can be extracted by adding a concentration dimension to the plot. A 'control map' based on this idea is popular in the geo-analytical community but could be more widely useful. In this map (it is not strictly a 'chart') $|d|$ is plotted directly against c, on linear or logarithmic axes as convenient. The control lines are functions of concentration and have to be plotted as such at $2\sigma_d$ and $3\sigma_d$. In addition, the median absolute difference is close in numerical value to σ_r, so adding this line to the map gives the analyst an extra test of the data, by counting the data points above and below this line — they should be roughly equal apart from random differences. The map is not a control chart because we are not presenting the data as a sequence. That usage is consistent with the concept of the run as an unchanging analytical system.

In some instances it may be preferable to use fitness-for-purpose uncertainty rather than repeatability standard deviation as a criterion of performance. The most likely circumstance is when the analyst has no previous knowledge of σ_d, such as might happen with determinations that are rarely undertaken. An example is discussed in the next section (§10.2).

Repeatability duplicates can be used to estimate the relationship between precision and the concentration of the analyte. However, a *large* number of results are needed to do this adequately.

Notes and further reading

- *If comparability within run is the customer's sole requirement then repeatability standard deviation can be used as the basis for uncertainty. The rationale for this usage lies in the definition of the measurand. In the circumstances referred to, the measurand is not the absolute concentration of the analyte, but the concentration differentials among two or more test materials analysed in the same run. If such results are released into the public domain, this limitation of the uncertainty should be made clear.*
- *For results that are normally distributed, the median of $|d|$ falls at $0.954\sigma_r$.*

10.2   Examples of Within-Run Quality Control

Key points
— A 'control map' is produced by plotting $|d|$ against concentration. Zero differences are set to a small positive value for plotting on logarithmic axes.
— Control lines are inserted at σ_r, $2\sigma_d$ and $3\sigma_d$.
— Overall, the precision of the results conforms well to the expected variation, and exceeds that of the external (10%) criterion, except possibly at low concentrations.

A large batch of samples of soil was analysed for selenium in one run in a completely random order. One in ten of the samples was analysed in duplicate. The duplicate results (ppm) are shown in the table. From previous experience, the repeatability standard deviation was expected to conform to the function $\sigma_r = \sqrt{0.015^2 + (0.04c)^2}$.

Se 1	Se 2	Se 1	Se 2
0.92	0.92	0.32	0.29
0.39	0.45	0.30	0.31
0.33	0.34	2.53	2.63
0.18	0.18	0.17	0.17
0.20	0.15	2.95	2.95
0.22	0.23	0.04	0.09
0.46	0.39	2.71	2.35
1.94	2.07	1.67	1.63
0.42	0.42	0.38	0.38
0.20	0.20	0.22	0.24
0.45	0.41	0.16	0.16

Differences between the pairs of results and corresponding values of σ_d were calculated. A dotplot of d/σ_d is shown in Fig. 10.2.1. The

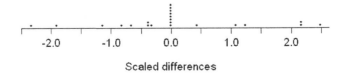

Scaled differences

Fig. 10.2.1.   Dotplot of scaled differences d/σ_d for the selenium data.

bunching of points at zero is the outcome of excessive rounding. Despite that, the data do not differ significantly from the standard normal distribution, so there is no evidence here that the original data deviated from expectations.

For further work, absolute differences $|d|$ were plotted against concentration. Control lines were calculated at σ_r, $2\sigma_d$ and $3\sigma_d$. The resulting map is shown in Fig. 10.2.2. It is perhaps easier to interpret the same features plotted on logarithmic axes (Fig. 10.2.3). (Differences of zero were set to 0.05 to allow them to be plotted on logarithmic axes: that does not affect the interpretation.) Thirteen points fall below the median line (dotted) and

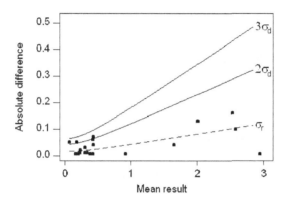

Fig. 10.2.2.   Control map for results (points) duplicated under repeatability conditions, with control lines at $3\sigma_d$ (solid), $2\sigma_d$ (solid) and σ_r (dashed).

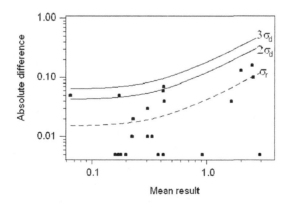

Fig. 10.2.3.   Same data and lines as Fig. 10.2.2, plotted on logarithmic axes.

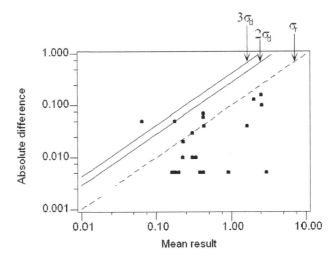

Fig. 10.2.4. The selenium results (points) plotted on a 10% control map. The control lines represent 10% RSDr.

nine above, indicating that, if anything, that the repeatability precision overall may be slightly better than expected. However, three points fall above the $2\sigma_d$ line, as compared with the expectation of 1.1. The probability of observing three or more points above this line, assuming the data conform to the expected repeatability, is $p = 0.02$. As these points are concentrated at lower concentrations, this suggests that the detection limit of the method was not as low as expected.

An alternative approach is to use an independent fitness-for-purpose criterion to judge the results. Figure 10.2.4 shows an absolute difference map with control lines for a repeatability relative standard deviation (RSDr) of 10%, i.e., for $\sigma_r/c = 0.1$. (This analytical precision would be sufficient for most environmental applications.) In this instance the results easily fulfil the requirement, with only four points above the theoretical median for 10% RSDr. There is one discrepant result (that is exceeding the $3\sigma_d$ line), at a low concentration, again suggesting that the detection limit may be higher than expected.

A singular advantage of using log-log plots with fixed RSDs is that general-purpose map blanks can be printed in advance in large numbers and the results quickly entered by hand. It is also relatively easy to write a macro that does the same job by computer.

Notes and further reading

- *The dataset is in file named **Selenium**.*
- *'A simple fitness-for-purpose control chart based on duplicate results obtained from routine test materials'. (2002). AMC Technical Briefs No. 9. Free download via www.rsc.org/amc.*

10.3   Internal Quality Control (IQC) and Run-to-Run Precision

Key points

— To apply statistical control to routine analysis, we have to use a surrogate, the control material, which resembles the test materials closely.
— Control limits are set according to the run-to-run precision for the system.

The purpose of internal quality control in analysis (IQC) is to ensure as far as possible that the magnitude of the errors affecting the analytical system is not changing during its routine use. The time-base for IQC is therefore the analytical run. During method validation we estimate the uncertainty of the method and show that it is fit for purpose. When the method is in use, every run of analysis should be checked to show that the errors of measurement are probably no larger than they were at validation time. For this purpose we employ the concept of statistical control, which means in general that some critical feature of the system is behaving like a normally-distributed variable. In industrial production, the critical feature is normally part of the specification, such as the length of a screw, and is readily available for measurement.

For chemical analysis, however, we have to generate separately a representative feature of the system. This is done by adding one or more 'control materials' to the run of test materials. The control materials are treated throughout in exactly the same manner as the test materials, from the weighing of the test portion to the final measurement. Clearly the control materials must be of the same type as the materials for which the analytical system was validated, in respect of matrix composition and analyte concentration. In that way the control materials act as a surrogate and their behaviour is a proper indicator of the performance of the system. The results obtained in successive runs can be plotted on a control chart (§10.4), which shows when the system becomes out of control and, by implication, needs investigation and possible remedial action before analysis resumes. Such a chart has to be set up with control lines determined by run-to-run precision.

When we undertake a number of successive runs in the same laboratory, the conditions of measurement will inevitably be different in each run: the instrument will be set up differently or a different instrument of the same type may be used. Newly prepared reagents or a new calibrator set may be used, perhaps by a different analyst. The environmental conditions in the laboratory may be different. Each run thus has a slightly different set of circumstances, giving rise to a 'run bias' effect. In the long term this variation looks like a random 'between-runs' effect in addition to the repeatability variation. Results replicated (that is, with the same material being analysed by the same method) in successive runs will therefore be more variable than repeatability replicates. This combined effect of repeatability and between-run variations is referred to here as run-to-run variation (and elsewhere, rather unhelpfully, as 'intermediate' variation). It is run-to-run standard deviation that should be used to set up control charts for internal quality control. An incorrect use of repeatability standard deviation for this purpose would result in too frequent an indication of loss of statistical control, whereas use of reproducibility standard deviation or standard uncertainty would result in too low a proportion.

It must be remembered that the parameters that define statistical control and used to set up control charts should refer only to the behaviour of the process itself. External criteria, such as certified reference values and their uncertainties are irrelevant to quality control *per se*. For the purposes of internal quality control, we need to know only whether the process (the analytical system) has *changed* since validation.

Notes

- *There is more information on control charts in §7.1.*
- *Thompson, M. and Wood, R. (1995). Harmonised Guidelines for Internal Quality Control in Analytical Chemistry Laboratories, Pure Appl. Chem.,* **67**, *pp. 649–666.*
- *'The J-chart: a simple plot that combines the capabilities of Shewhart and cusum charts, for use in analytical quality control'. (2003). AMC Technical Briefs No. 12. Free download via www.rsc.org/amc.*

10.4 Setting Up a Control Chart

> **Key points**
> — It is difficult to estimate run-to-run standard deviation accurately unless routine conditions prevail during the measurements.
> — Early results may be atypically disperse.
> — An interim chart can be set up immediately after validation by using a repeatability mean and an inflated repeatability precision.
> — The interim chart should be replaced after ten or more runs with run-to-run statistics and reviewed periodically after that.

There are practical problems in setting up a control system for analysis. Run-to run standard deviation cannot be estimated adequately in the usual type of one-off validation. Real-life replication is required, over an extended series of runs. To achieve this realism, any control material has to be in a random position in a run-length sequence of typical test materials. Many observations are needed to estimate the standard deviation with suitable accuracy, far more than the usual ten. These conditions cannot be realised except during routine use of the method. Moreover, the analysts will not be familiar with the method at initial validation time, and will produce results of atypically low precision: almost invariably an improved precision comes with experience of the method.

How then does the analyst actually start the control chart? The best strategy is to use an interim control chart and update it as more information becomes available, as follows.

- Start an interim control chart with the mean result for the control material established at validation time. Define the control limits on the basis of 1.6 times a repeatability standard deviation estimated from the results at validation. (The factor of 1.6 is derived from a broadly-applicable empirical observation of the magnitude of run-to-run variation.) Do *not* use uncertainty values attached to a certified value of a reference material: that is a description of knowledge about the material, not about the analytical system.

- After results have accumulated from ten runs, replace the control limits with those based on the robust estimates of the mean and standard deviation of the new results. After further (say about 30) results have accumulated these estimates should be checked.

- Review the control limits occasionally, as may be necessary if the mean of the process has clearly changed. This requires the exercise of judgement and should not be done without careful consideration. A substantive change may demand a partial revalidation of the process.

Notes and further reading

- Howarth, R. J. (1995). *Quality Control Charting for the Analytical Laboratory. Part 1. Univariate Methods. A Review*, Analyst, **120**, pp. 1851–1873.
- *Internal quality control in routine analysis.* (2010). AMC Technical Briefs No. 46. Free download via www.rsc.org/amc.

10.5   Internal Quality Control — Example

> **Key points**
> — The best option for a control chart for this dataset is with control lines based on robust statistics of more than about 15 runs of analysis.
> — The interim chart based on inflating an estimate of repeatability standard deviation was very similar to the 'permanent' control chart based on run-to-run precision.

Here we consider some quality control data derived from the routine analysis of soil samples by inductively-coupled plasma atomic emission spectrometry. The element of interest is zinc, and the main emphasis is on the typical (but perhaps unexpected) difficulty in determining the control limits. Note that we do not proceed on the assumption that the data will be exactly normally distributed. A more likely occurrence would be that the majority of points are roughly normally distributed and a minority are outliers.

Four methods are illustrated in Fig. 10.5.1. Graph (a) shows the mean and action limits, determined from rolling statistics (simple mean and standard deviation), from each set of ten successive results from 50 runs. (For example, the limits at Run 50 are based on data from Runs 41–50.) The positions of the control lines are very variable, showing that any choice of just ten runs would give rise to highly questionable outcomes. Graph (b) shows the corresponding control lines determined from all data up to the current round (that is, simple statistics based on runs 1 to n). The lines

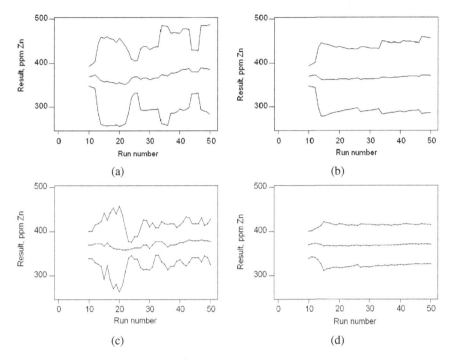

Fig. 10.5.1. Positions of control lines (mean and action limits) estimated by various methods at different points in the accumulation of data. (a) Rolling groups of ten data, simple statistics. (b) All data up to run n, simple statistics. (c) Rolling groups of ten data, robust statistics. (d) All data up to run n, robust statistics.

are more stable in position after Run 14, but noticeably affected by some following individual runs. Graph (c) shows the statistics estimated by a robust method (Huber's proposal H15, see §7.6) with rolling groups of ten successive results. The lines are very unstable until Run 25, and after that they are narrower but still somewhat ragged. When the robust statistics are calculated from all data up to Run n (Graph d), the resulting lines are narrow and stable almost from the start. In this example at least, robust statistics from a long succession of data would give the best outcome for a control chart. Using statistics from data from the first run up to any point after run 20 would give a serviceable control chart.

For setting up an interim control chart, the initial repeatability statistics were a mean of 370 and a standard deviation of 8.9 ppm. The interim estimate of run-to-run standard deviation was therefore $1.6 \times 8.9 = 14.2$ (see §10.4). A chart with control lines based on $370 \pm k \times 14.2$, $k = 2, 3$ was

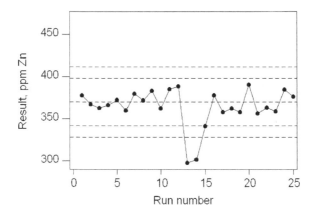

Fig. 10.5.2.   Interim control chart based on a standard deviation of 1.6 times the repeatability value.

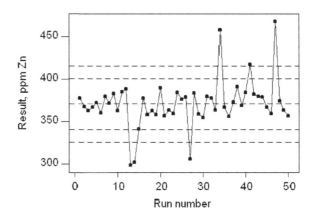

Fig. 10.5.3.   'Permanent' control chart for zinc with control limits based on robust statistics from the whole dataset, namely a mean of 370.6 and a standard deviation of 14.8.

used as the interim chart. Figure 10.5.2 shows this chart used for the first 25 rounds. The chart has the appearance of well-behaved data apart from Runs 13 and 14, which are out of control.

A 'permanent' control chart based on option d (all data up to Run 50, robust statistics) is shown in Fig. 10.5.3. It is remarkably similar to the interim chart in this instance. There are a number of clear outliers signifying out-of-control conditions (Runs 13, 14, 27, 34 and 47), and one marginal case (Run 41). This has every appearance of a sensible control chart.

Notes and further reading

- *The data for this example were a subset from file **AGRG-IQC** downloadable from AMC Datasets via www.rsc.org/amc. This dataset comprises multielement results from 156 successive runs (including rejected runs), with a variable number of repeat results for each run. The subset used here comprised the first result for zinc from each run up to Run 50. The repeatability standard deviation for the interim chart was obtained independently.*
- *The closeness of agreement between the control lines of the interim and permanent control charts was in this instance unusually good.*

10.6 Multivariate Internal Quality Control

Key points

— Results from simultaneous multi-analyte methods are likely to be correlated because of variations in procedure that affect all analytes.
— Such correlations may be diagnostic and point to specific causes of problems.
— Multiple symbolic control charts are useful because they clearly show runs where many channels are affected simultaneously and analytes that are affected in many runs.

Chemical analysis with multiple outputs, either simultaneous (e.g., ICPAES) or nearly simultaneous (e.g., HPLC) is commonplace nowadays. The question of how best to apply IQC principles to such systems is often discussed. The use of multivariate statistical methods in this context has seldom been reported so far, and these methods are beyond the scope of this book. In any event, such methods may provide a robust account of the analytical system as a whole, but both the analyst and the customer will require information about the validity of results for each individual analyte. The present discussion is therefore limited to the multiple use of univariate methods.

One preliminary consideration is the extent to which the variables (the results for the analytes) are correlated. Variation in parts of the analytical procedure that are common to all analytes will tend to cause correlationcorrelation. Variation in the volume of test solution injected into a chromatograph would be of that kind: it would tend to affect all analytes in proportion. Other variations in method may have outcomes that are limited to a subset of analytes. In an acid decomposition of a soil sample, variation in the final temperature of drying would cause variation in the recovery of a suite of volatile elements (e.g., Hg, As, Se) while other elements would be completely unaffected. An extreme example with only one analyte affected could be caused by malfunction in a single channel of a multichannel instrument. These features may be worth some consideration for their potential diagnostic value in instances of out-of-control runs.

Another aspect of this is the well-known 'Bonferroni' problem. Suppose we have a system measuring 30 analytes simultaneously, and we are considering results that fall outside the warning limits of a control chart (that is, the approximate 95% confidence limits). If all of the channels were independent (not correlated), in every run we would expect to see results for one or more analytes in the warning zone purely by chance. Isolated instances almost certainly would mean nothing. The probability (under the usual assumptions) of observing a result for a single analyte outside the action limits is about one in 300, which is so rare in an in-control system that we are justified in assuming that the system has changed. But in an uncorrelated 30-channel system operating in control we could expect to encounter a result outside the action limits with a probability of one in ten. That might lead the unwary to reject data at a quite unnecessary rate.

Fortunately, largely uncorrelated multi-analyte systems are seldom encountered. When things go wrong the effects tend to be visible on a number of analytes simultaneously. Such occurrences are clearly visible in a multiple symbolic control chart. To make such a chart the results x for each channel are standardised as $z = (x - \hat{\mu})/\hat{\sigma}$, where $\hat{\mu}$, $\hat{\sigma}$ are respectively the estimated mean and standard deviation for that channel. The values of z are then converted into symbols indicating the zone into which the result would fall on the corresponding Shewhart chart. The symbols are then plotted according to analyte (rows) and run number (columns).

Figure 10.6.1 shows such a chart for the results of 25 elements in a control material over the first 50 runs of a routine procedure. (The example used previously [§10.5.3] represents one row of this data.) On this chart the great majority of instances of results falling outside action limits are concentrated into a few runs: a large number of instances occur together (and usually in the same direction), pre-eminently in runs 13, 14, 15, 24, 27, 34, 41 and 47. (Another phenomenon potentially present in multiple symbolic charts is where a particular *analyte* shows signs of persistent problems

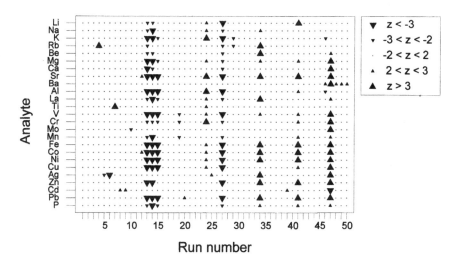

Fig. 10.6.1. Symbolic multiple-analyte control chart.

over a number of runs. That would be indicative of a specific problem with that element. No example is present in Fig. 10.6.1.)

It is clear that the channels in this example are mostly highly correlated. Even when the anomalous runs were deleted from the dataset, there is still a degree of correlation among the variations. The following correlations are typical of the whole correlation matrix.

	Li	Na	K	Rb
Li	1.0	0.5	0.5	0.3
Na	0.5	1.0	0.9	0.4
K	0.5	0.9	1.0	0.4
Rb	0.3	0.4	0.4	1.0

Notes and further reading

- When a number of analytes in a material are determined by quite separate methods on separate test portions, the results can be assumed to be independent.

- The data for this example were a subset from file named **AGRG-IQC**, downloadable from AMC Datasets on the website www.rsc.org/amc. This dataset comprises multielement results from 156 successive runs (including rejected runs), with a variable number of repeat results for each run.

Chapter 11

Proficiency Testing

Proficiency testing, an externally-provided means for participants to check the accuracy of their routine measurements, is now incumbent on all laboratories seeking accreditation. The conversion of participants' results into meaningful and readily understandable scores is almost universal. Nearly all such scoring systems are based on the properties of the normal distribution, but some statistical methods needing special software may be required in the process. While it is the job of the scheme provider to execute these methods, it is useful for the participant to be aware of what is involved.

11.1 Proficiency Tests — Purpose and Organisation

> **Key points**
> — Proficiency tests are regular interlaboratory comparisons of results obtained by 'blind' analysis.
> — The main purpose is to help laboratories to achieve a suitably low uncertainty.
> — Participation in a scheme (where available) is an almost universal requirement for accreditation.
> — Participants' results are usually converted into scores that give an indication of accuracy.

Proficiency tests are interlaboratory exercises, provided on a regular basis, to allow participating laboratories to check the accuracy of their results. For each round of the scheme, the scheme provider sends portions of one or more test materials to the participants, who analyse the materials 'blind' (that is, with no indication of the concentrations of the analytes) by their routine methods. The materials and analytes should be typical of the participants' normal work. The materials should be sufficiently close to homogeneous and stable, so that variations among the results reflect accurately the variations in the participants' performance rather than variations in the test material. After the reporting deadline, the provider processes the results, usually converting them into scores that gives an indication of the accuracy, often in relation to a predetermined maximum level of uncertainty. The provider sends a report of the round to the participants, showing the results and/or scores of all participants.

Proficiency tests rounds are provided at various frequencies, most commonly several times per year. They cannot therefore act as a substitute for internal quality control (§10.3), which should be carried out in every run of analysis.

The primary purpose of proficiency testing is to enable participants to be confident about their normal analytical methods. If an unexpected inaccuracy in their routine results is detected, an investigation can be triggered and remedial actions instituted where necessary. This function is so important that participation in a proficiency test, where one is available, has been made a universal requirement for accreditation. Moreover, accreditation agencies expect participants to have and apply a written procedure for dealing with poor scores. However, accreditation has had the unfortunate effect of encouraging participants to try to excel in accuracy rather than merely to assess the performance of routine operations. This tendency is reinforced when laboratories use their scores in promotional activities, for example by quoting favourable scores in tenders for work, or for monitoring the performance of individual analysts. These secondary uses have to a certain extent subverted the original ethos of proficiency testing.

Notes and further reading

- Thompson, M., Ellison, S. L. R. and Wood, R. (2006). The Internaional Harmonised Protocol for the Proficiency Testing of Analytical Chemistry Laboratories, Pure Appl. Chem., **78**, pp. 145–196.(Free download from IUPAC website.)
- ISO Guide 43. Proficiency testing by interlaboratory comparisons – Part 1: development and operation of proficiency testing schemes. (1994). International Organisation for Standardisation, Geneva.
- Lawn, R. E., Thompson, M. and Walker, R. F. (1997). Proficiency Testing in Analytical Chemistry, The Royal Society of Chemistry, Cambridge.

- *ISO 13528. Statistical methods for use in proficiency testing by interlaboratory comparisons. (2005). International Organisation for Standardisation, Geneva.*
- *ILAC-G13. Guidelines for the requirements for the competence of providers of proficiency testing schemes. (2000). (Available free online at www.ilac.org/documents/)*

11.2 Scoring in Proficiency Tests

Key points

— Converting a participant's result into a score is pointless unless it adds information about the accuracy of the result.
— An ideal score should convey the same information about accuracy regardless of the nature of the analytical measurement.
— The z-score is close to ideal.

A participant's result x for an analyte in a round of a proficiency test is usually converted into a score that reflects the accuracy of the results. The ideal score should be universally applicable: a particular value, say 1.5, should convey the same information about the accuracy of a result, regardless of the analyte, its concentration, the test material or the physical principle underlying the measurement. In fact, scoring is pointless unless it has this property. The z-score, given by $z = (x - x_A)/\sigma_p$, is appropriate, where the 'assigned value' x_A is the scheme provider's best estimate of the true value of the measurand, and σ_p is the 'standard deviation for proficiency' (also known informally as the 'target value'). However, the efficacy of a scheme depends critically on the selection of appropriate values for x_A and σ_p.

A hypothetical laboratory using an unbiased method producing results with an uncertainty $u = \sigma_p$ would tend to produce z-scores that are a random sample from a standard normal distribution $N(0,1)$, that is, with a mean of zero and a variance of unity. Consequently, it is appropriate to interpret z-scores on this basis, as we would expect about 95% of z-scores from exactly compliant laboratories to fall between ± 2 and very few to fall outside ± 3. Real-life laboratories will not be exactly compliant, however. Laboratories operating with poor uncertainty ($u > \sigma_p$) and/or with a bias tend to produce higher proportions of scores outside these limits. In contrast, laboratories operating with no bias and an uncertainty

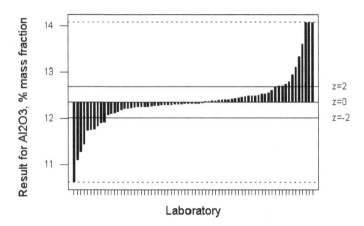

Fig. 11.2.1. Barchart of results for alumina showing an approximately symmetrical distribution in a round of the GeoPT proficiency test.

smaller than σ_p tend to produce a smaller proportion of scores outside the limits.

Some examples of results and z-scores obtained in rounds of some proficiency tests are shown in Figs. 11.2.1, 11.4.2 and 11.4.4. In typical reports, the participants' scores are shown as a bar chart or ordered bar chart, with the individual laboratory identified by an anonymised code. In some instances the number of participants is so large that a bar chart is impracticable and is replaced by a histogram.

11.3   Setting the Value of the Assigned Value x_A in Proficiency Tests

> **Key points**
> — There are a number of possible ways of determining an 'assigned value'.
> — The value most often used is the consensus of the participants' results.

There are several possibilities for the choice of assigned value x_A.

- *The certified value* of the analyte in a certified reference material (CRM). Metrologically sound, this value is seldom used because the cost of a CRM is usually too great for use in a routine manner. In addition, the uncertainty of the certified value is often too large to be useful.

- *A value from a national reference laboratory* obtained by a method such as isotope dilution mass spectrometry. Similar comments apply to this value.

- *A value obtained by analysis* alongside a number of matrix-matched CRMs in the same analytical run (that is, using the CRMs as calibrators). There are seldom enough appropriate CRMs available.

- *A value based on formulation.* This can be used where the analyte is added gravimetrically or volumetrically to a base material containing none. Sometimes applicable, there are often difficulties in accurately spiking the base material with low trace levels of analyte.

- *A consensus of expert laboratories.* One problem with this assigned value stems from the difficulty of identifying the expert laboratories to the satisfaction of every stakeholder. A practical problem is that the variation between the experts' results is often comparable with that of the whole participant set, and the assigned value does not have a sufficiently small uncertainty.

- *A consensus of all participants.* This is by far the most commonly used assigned value, and it costs nothing. A consensus is usually easy to identify and has a sufficiently small standard error if there are more than about 20 participants. The consensus has been criticised on metrological grounds, as it is perfectly possible for the great majority of the participants to be using a biased analytical method. In such instances, which are occasionally detected, there would be a latent uncertainty in the assigned value, and participants using an unbiased method could receive poor z-scores. However, at present there is seldom an economically viable alternative. Methods of determining a consensus are considered below (§11.4).

Further reading

- *Thompson, M., Ellison, S. L. R. and Wood, R. (2006). The Internaional Harmonised Protocol for the Proficiency Testing of Analytical Chemistry Laboratories, Pure Appl. Chem., 78, pp. 145–196. (Free download from IUPAC website.)*

11.4   Calculating a Participant Consensus

Key points
— A robust mean is usually a good estimator of a consensus if the results from a round of a proficiency test seem unimodal and (outliers aside) roughly symmetrical.
— If the dataset seems to be unimodal but skewed, a mode estimated by kernel density methods may be a suitable consensus.
— Where the results are apparently multimodal, it may be impossible to find a consensus.

In the context of proficiency testing, 'consensus' does not mean absolute concordance, but an identifiable and unique point of maximum agreement between the participants' results. All of the usual measures of central tendency have been considered in this context. In addition to the value of the selected statistic itself, an estimate of its uncertainty is required to ensure that the assigned value is sufficiently stable. Methods for estimating these statistics abound, but experience and judgement are needed to select the method appropriate for particular datasets.

- *The mean.* The almost inevitable presence of outliers and heavy tails in sets of results from proficiency tests means that the arithmetic mean may be biased and the variance inflated. One of several robust estimates is suitable to avoid these problems if the dataset is unimodal and reasonably close to symmetric (e.g., Fig. 11.4.1). In such datasets the various estimates of central tendency are almost coincident, and the robust mean is a good estimator. The standard error of the robust mean can be estimated as $\hat{\sigma}_{rob}/\sqrt{n}$ from the robust standard deviation $\hat{\sigma}_{rob}$ and the number of participants n. (This is a reasonable estimate if the robustification is not too severe: otherwise the value of n should be adjusted for any down-weighting.)
- *The median.* The median is a type of robust mean but is more resistant than some estimators to the influence of skewness, which may appear in proficiency tests datasets through the use of a number of methods with differing detection limits. However, the mode is usually preferable for skewed distributions.

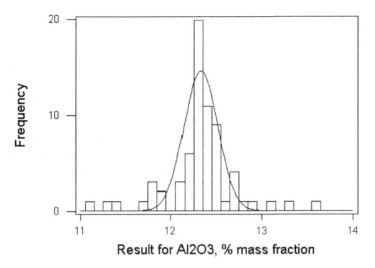

Fig. 11.4.1. Histogram of results for alumina in a rock test material from a round of the GeoPT proficiency test. Extreme outliers have been omitted. The results are heavy tailed in comparison with the robustly fitted normal distribution (solid line). Same data as Fig. 11.2.1.

- *The mode.* The mode is intuitively attractive as a consensus estimator, and serves well even if the dataset shows a moderate degree of skewness, e.g., Fig. 11.4.2. The mode of a smooth distribution is the point of highest density. Real datasets presented as histograms or dotplots are not smooth, however, owing to the class boundaries or because of the limited digit resolution of the data. A degree of smoothing is therefore required to identify the mode, and to check that there is indeed only one mode. This smoothing can be conveniently carried out by kernel density estimation (Fig. 11.4.3). The standard error of the mode can be estimated via the bootstrap (a computer-intensive method of estimation).

In some instances more than one mode (other than outliers) may be apparent (Figs. 11.4.4–11.4.6). This could happen if substantial subsets of the participants used one of several discrepant analytical methods or variants of a single method. In such instances it is usually not possible to identify a consensus. Occasionally, however, there may be additional evidence (such as use by participants of a prescribed method) that enables the provider to determine that one such mode represents reliable results and other modes suspect results. In that case the provider can, with due

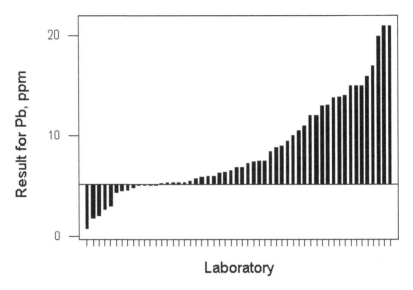

Fig. 11.4.2.   Barchart of results for lead, from the GeoPT proficiency test, showing a strong positive skew.

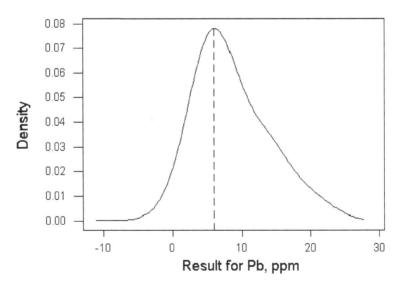

Fig. 11.4.3.   Kernel density representation of results for lead from a round of the GeoPT proficiency test, showing a positive skew and the single mode at about 6 ppm. (Same data as Fig. 11.4.2.) (The sub-zero density in the low tail is the result of smoothing.)

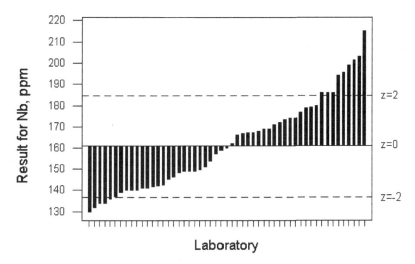

Fig. 11.4.4. Bar chart of results for niobium from a round of the GeoPT proficiency test. Despite the high proportion of results with good z-scores, there is a suggestion of multimodality.

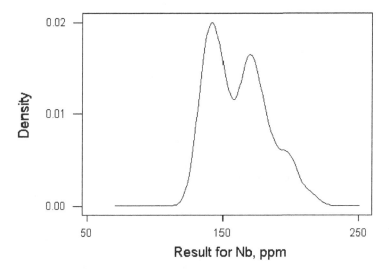

Fig. 11.4.5. Kernel density representation of results for niobium from a round of the GeoPT proficiency test (same data as Fig. 11.4.4). The tendency to multimodality is clear.

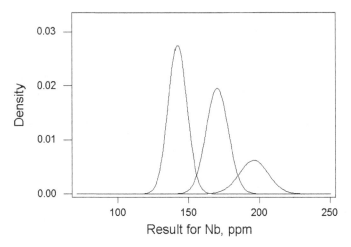

Fig. 11.4.6.   The results for niobium from a round of the GeoPT proficiency test are well described as a mixture model of three normally distributed subsets. (Same data as Fig. 11.4.4.)

caution, use the reliable mode as the consensus. A mixture model is a useful additional technique for characterising such subsets of the data.

Notes and further reading

- *Some statistical methods mentioned in this section are beyond the scope of this book, but straightforward accounts can be read in the AMC Technical Briefs listed below. A kernel density provides a smooth representation of data by replacing data points by small (usually) normal distributions and then adding the resulting densities at each point on the measurement axis. The bootstrap estimates distributions of statistics by resampling the data with replacement a large number of times. A mixture model regards the dataset as a mixture of random samples of observations from two or more different distributions.*

- *'Representing data distributions with kernel density estimates'. (Revised March 2006). AMC Technical Briefs No. 4. Free download from www.rsc.org/amc.*

- *'The bootstrap: a simple approach to estimating standard errors and confidence intervals when theory fails'. (August 2001). AMC Technical Briefs No. 8. Free download from www.rsc.org/amc.*

- *'Mixture models for describing multimodal data'. (March 2006). AMC Technical Briefs No. 23. Free download from www.rsc.org/amc.*

11.5 Setting the Value of the 'Target Value' σ_p in Proficiency Tests

> **Key points**
> — In best practice the target value should be a criterion known to participants in advance and describing the uncertainty regarded as fit for purpose in the relevant application sector.
> — Target values based on a robustified standard deviation of the participants' results do not provide any useful additional information.

The best value of σ_p is simply the uncertainty that is regarded as fit for purpose in the application sector. That information should be available to the participants before the analysis. Under this convention z-scores will be comparable over any analytical method, analyte or matrix. For example, a z-score outside the range ± 3 will always show that the original result was not fit for purpose. It is important to emphasise (because it is widely misunderstood) that σ_p is not intended to predict the uncertainty of individual laboratories. Equally it does not imply that the collected z-scores of all of the participants will be a random sample from the standard normal deviation $N(0,1)$. Individual laboratories will tend to have different precisions and biases that jointly contribute to the between-laboratory variation. Thus real datasets deviate from $N(0,1)$ to a greater or lesser extent, often with heavy tails, a proportion of outliers and, occasionally, skews or multiple modes. The value of σ_p is not intended to predict that diversity: rather it is set to prescribe in advance the uncertainty that is required by the scheme provider. It is up to the participants to attempt to comply. The z-scores then give a good idea of the degree of compliance. If the fit-for-purpose uncertainty varies with the concentration of the analyte, it should be expressed as a functional relationship.

Some proficiency tests use a robust standard deviation of the participants' results in a round to define σ_p. The resulting z-scores will nearly always show in excess of 90% laboratories with scores between ± 2. This may be comforting for the participants, the great majority of whom could claim that their result was 'satisfactory', and equally so for the provider, who could claim that the scheme was achieving its purpose. In fact such a score tells us nothing about whether the results are accurate enough in relation to their intended purpose. Moreover, it does not tend to reduce the variation among the great majority of the participant laboratories. Finally, it does not give the participant any prior guidance as to the standard of accuracy that is required: such guidance is necessary so that appropriate analytical methods can be chosen in advance or modified to meet the requirement. In short, scoring is pointless unless it adds information to that already present in the raw results.

11.6   Using Information from Proficiency Test Scores

> **Key points**
> — z-scores should be regarded as action limits rather than a method of classifying participants.
> — Laboratories should have a documented strategy for examining and acting upon z-scores.

In a scheme where the σ_p value is determined by fitness for purpose, z-scores in the range ± 2 can be regarded as calling for no action by the participant. Scores outside the range ± 3 would be very unusual for a participant conforming to the fitness criterion and can be taken as calling for investigation and possibly remedial action, such as modification of the analytical system. Scores in the intermediate range would not be especially uncommon and isolated instances could be ignored. So $z = \pm 3$ can best be regarded as defining action limits.

There is a temptation amongst practitioners to regard these arbitrary limits as class boundaries and to name the classes accordingly, for instance, 'satisfactory' for $|z| < 2$ and 'unsatisfactory' for $|z| > 3$. These class labels are best avoided. There is a danger that they will be interpreted literally and misapplied, especially by those not familiar with statistical inference. There is also a tendency among non-scientists to want to construct a ranked 'league table' from a set of z-scores. That is especially invidious, and should be strongly discouraged if encountered, as ranks can change greatly from round to round without any underlying change in the performances of the participants.

There is also an understandable desire for people to try to summarise performance by averaging z-scores, for a single analyte over a period of time or for a number of different analytes at a particular time. There are several problems associated with creating these summary scores. One such is that a single z-score of large magnitude will have a persistent effect in time, possibly long after the problem giving rise to it has been rectified. Another is that, in an average score from a number of analytes, a particular analyte may consistently attract a score of large magnitude that is hidden in the average. Finally, when several analytes are determined 'simultaneously', the results are unlikely to be independent and the average misleading (unless corrected for covariance).

No such problems are attached to interpreting z-scores by standard univariate control chart procedures, using the usual rules of interpretation (§7.1). Either Shewhart charts or zone charts are suitable for this purpose. In Fig. 11.6.1 we see no trend in the scores for an analyte in successive rounds, but the score in Round 10 is less than -3 so the analytical system needs investigating. If several analytes are determined together in successive rounds, parallel symbolic control charts are especially informative. In Fig. 11.6.2 we see five instances where $|z| > 3$, each calling for investigation. Under the usual rules, Analyte 2 in Round 11 would also trigger investigation, because there are two successive results where $-3 < z < -2$. We can also see that Analyte 2 alone is unduly prone to low results while, in Round 4, five out of the six analytes attract unduly

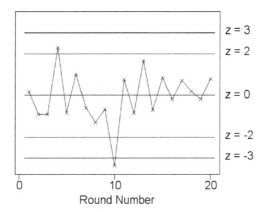

Fig. 11.6.1.   z-scores for a single analyte from successive rounds of a proficiency test.

Multiple z-score chart

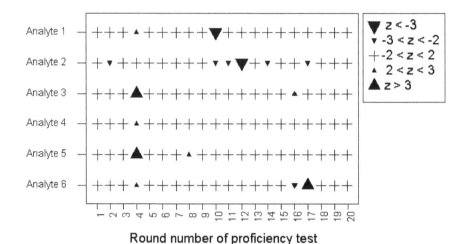

Round number of proficiency test

Fig. 11.6.2. Multiple univariate control chart used to summarise the results for a number of analytes determined in successive rounds of a proficiency test. (Analyte 1 results are also shown in Fig. 11.6.1.)

high results, possibly by a procedural mistake affecting all of them in the same way.

Notes and further reading

- *A z-score of unexpectedly large magnitude shows that the internal quality control system also needs investigation: a problem causing the troublesome z-score (unless due to a sporadic mistake) should have been detected promptly by internal quality control and the result rejected by the analyst.*
- *It is important to realise that a poorly-performing participant can still receive a majority of 'satisfactory' z-scores. If a participant's method was unbiased but the standard deviation were twice the target value (i.e., $2\sigma_p$) the laboratory would on average still receive a z-score between ± 2 on about 67% of occasions and between ± 3 on about 87% of occasions.*
- *'Understanding and acting on scores obtained in proficiency testing schemes'. (December 2002). AMC Technical Briefs No. 11. Free download from www.rsc.org/amc.*

11.7 Occasional Use of Certified Reference Materials in Quality Assurance

Key points

— Certified reference materials (CRMs) are not ideal for internal quality control. For several reasons, despite their providing a route to traceability.

— CRMs as an occasional check on accuracy are best regarded as akin to proficiency testing.

Certified reference materials (CRMs) are sometimes advocated for use as control materials in internal quality control (IQC) of analysis, in that the CRM is a direct route to an acceptable traceability. In many ways, however, it is better to keep the concepts of quality control and traceability distinct. The use of a CRM (where available) on a scale appropriate for IQC would usually be too expensive but on a lesser scale ineffectual. Moreover, there is a discrepancy in concept between IQC and the CRM. The control chart is based on the properties (i.e., the mean and variance) of the whole analytical system. For the CRM, in contrast, the certified value and its uncertainty describe the material alone.

The analysis of a CRM, however, can provide a useful occasional check of an ongoing analytical system, if a sufficiently close match to the test materials can be found. In such instances it is better to regard the action as a kind of one-laboratory proficiency test rather than part of IQC. The outcome could be assessed by calculating a 'pseudo-z-score' $z' = (x - x_{crm}) / \sqrt{u_f^2 + u_{crm}^2}$ from the result x, the uncertainty u_f regarded as fit for purpose for the result, the certified value x_{crm} and its uncertainty u_{crm}. The fit-for-purpose uncertainty u_f would have to be specified in advance, possibly as a function of concentration. It is very important to notice that any such test would be pointless unless the uncertainty on the certified value is negligible. Unless $u_{crm} <\approx u_f/2$, the score z' would reflect the uncertainty in the certified value to an undue extent and mask the behaviour of the analytical system.

Chapter 12

Sampling in Chemical Measurement

This chapter is concerned with the statistics involved in the neglected topic of uncertainty from sampling. In many application sectors sampling uncertainty is a substantial or even dominant term in the uncertainty budget and therefore very important. The end-user of analytical results needs to know the combined uncertainty (analytical plus sampling) to make valid decisions about the sampling 'target', decisions such as the commercial value of a batch or lot of a material, or whether material conforms with a legal or contractual specification.

12.1   Traditional and Modern Approaches to Sampling Uncertainty

---

Key points

— Sampling uncertainty is traditionally ignored if the sample is 'representative'.

— The modern approach regards sampling as an integral part of the measurement process and includes its contribution to the combined uncertainty.

— Analytical chemists should use the recommended terminology of sampling.

Nearly all analysis is preceded by sampling, the process of taking a small portion (the sample) from the much larger amount (the target), the composition of which is in question. The sample is small enough to be removed to the laboratory for further physical preparation such as grinding before analysis, while the target usually is not. Of course, taking a sample is pointless unless it reasonably approximates the average composition of the target: such a sample is called 'representative'. Obtaining a representative sample is sometimes very difficult, especially when the target is very large, a shipload of ore for example, and parts of the target are difficult to access. Procedures for obtaining representative samples of nearly all materials of commerce have been produced and often form parts of contractual agreements or legal requirements. Such samples are usually accepted by analytical chemists and end-users of analytical data without further consideration. In effect, the uncertainty introduced into the final analytical result by sampling is ignored in the traditional approach.

A more modern approach to sampling avoids the idea of a representative sample, but considers the uncertainty introduced by the sampling as an integral part of the measurement uncertainty. This approach provides a quantitative basis for the often repeated saying among analytical chemists, that 'the result is only as good as the sample'. It is an important development because the end user of analytical results needs information about the target, not the sample. Moreover, we cannot ensure that we are using resources optimally unless we can compare the uncertainty contributions from sampling and analysis. Despite the modern trend towards considering sampling and analysis as parts of a single measurement operation, their contributions usually have to be estimated separately because sampling is often conducted at a location remote from the analytical laboratory and seldom under the complete control of analytical chemists.

The following sections in this chapter refer to the estimation and use of uncertainty from sampling. They do not provide instructions for obtaining representative samples of specific materials. However, many textbooks refer to the general principles of sampling for chemical analysis.

The word 'sample' is often used informally among analytical chemists to indicate 'analyte', 'test portion', 'test material', 'test solution', 'aliquot', 'matrix', and so on. Such usage should be discontinued to avoid confusion. The recommended terminology for the various stages of sampling is shown here. 'Sampling' is usually taken to include all operations down to the preparation of the test sample. 'Analysis' is taken to mean all of the subsequent steps, starting with the selection and weighing of the test portion. Any residual heterogeneity in the test sample gives rise to an uncertainty that is attributed to the analytical variation. Some stages (subsample, laboratory sample) are omitted or merged in many instances. Key terms are underlined.

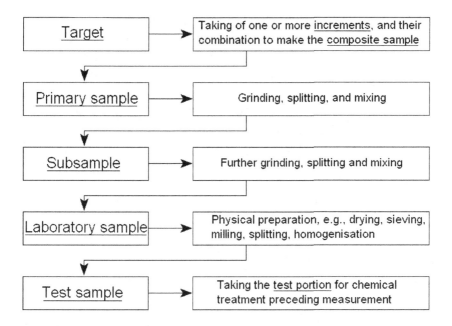

Notes and further reading

- *'Terminology – the key to understanding analytical science. Part 2: Sampling and sample preparation'. (March 2005). AMC Technical Briefs No. 19.*
- *Crosby, N.T. and Patel, I. (1995). General Principles of Good Sampling Practice, Royal Society of Chemistry, Cambridge.*

12.2   Sampling Uncertainty in Context

Key points

— Only the combined uncertainty (sampling plus analytical) is relevant to the customer's needs.

— Taking proper account of sampling uncertainty can affect decisions based on analysis.

— The concepts of bias and precision can be applied to sampling.

All sampling targets are actually or potentially heterogeneous: the chemical composition can vary from point to point in the material. This implies that replicate test samples from a single target will differ in composition from each other and from the target. This variation gives rise to uncertainty from sampling u_s that is additional to and independent of the uncertainty u_a derived from purely analytical activities. The combined uncertainty on the composition of the target is thus $u = \sqrt{u_s^2 + u_a^2}$. It is this combined uncertainty that is relevant to the needs of the end-user of the data, who is required to make rational decisions about the target (not about the laboratory sample). Typical decisions are the probable commercial value of a batch of material, or whether it is probably within specification. In a high proportion of instances, u_s makes a substantial contribution to the combined uncertainty. In environmental studies, and in the examination of raw materials such as food or ores, u_s could even be the dominant contribution. This fact has important implications for decision making by end-users of analytical data, legislators, enforcers of legislation and analytical chemists (§8.3).

Analysis is often conducted to determine whether the material in a target conforms to a legislative or contractual limit, either an upper limit or a lower limit. Hitherto, only the analytical uncertainty was taken into account in making such decisions. The sampling uncertainty was in effect taken to be zero. That basis for decision making is illustrated in Fig. 12.2.1 as examples A and A'. As the analytical uncertainties do not encompass the limit value, the results are regarded as definitive. Result A is clearly below the limit and result A' above. That, however, is not a logical standpoint unless the sample is defined as the target by law. (For instance, a single bottle of milk may have to conform to a regulation, rather than the

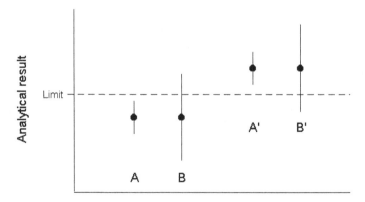

Fig. 12.2.1.   Measurement results (points) and expanded uncertainties (vertical bars) in relation to a decision limit: A, A' when only analytical uncertainty is considered; B, B' when combined uncertainty (analytical plus sampling) is considered.

whole consignment from which the bottle is an example.) The uncertainty from sampling should be taken into proper account in decision making. If sampling uncertainty is taken into account, so that the combined uncertainty is larger (B, B′), the same analytical result would give rise to a different decision.

To create a conceptual framework for sampling uncertainty, we must consider ideas like precision and bias applied to sampling, and carry out operations such as the validation of methods and quality control. These terms are familiar when applied to analytical methods but in sampling there are differences in the way that they can be tackled. This is because sampling uncertainty is partly the outcome of the heterogeneity of the material under test as well as the process of collecting it. Moreover, successive targets of the same type of material can vary in the degree of heterogeneity, so that the value of u_s could vary from target to target. A working value of u_s will have to be a robust average that is typical of the material as a whole. This potential variation in the degree of heterogeneity implies that quality control is especially important in sampling. An analytical result on a sample from a target with an anomalously high value of u_s could be unfit for purpose, even though the validated sampling protocol was scrupulously followed.

Notes and further reading

- *At the time of writing, the subject of sampling uncertainty is immature: to date very little practical experience has accrued in using tools for dealing with uncertainty from sampling, and even less has been documented.*
- *'What is uncertainty from sampling, and why is it important?' (July 2008). AMC Technical Brief No. 16A. Free download via www.rsc.org/amc.*
- *Ramsey, M. H. and Thompson, M. (2007). Sampling Uncertainty in the Context of Fitness for Purpose, Accred. Qual. Assur., 12, pp. 503–513.*
- *Ramsey, M. H. and Ellison, S. L. R. (eds). (2007). Measurement Uncertainty Arising from Sampling — a Guide to Methods and Approaches. The Guide is the joint production, under the Chairmanship of Prof. M. H. Ramsey, of Eurachem, CITAC, Eurolab, Nordtest and the Analytical Methods Committee. It contains chapters on fundamental concepts, estimation of sampling uncertainty and management issues. Six practical examples are examined in detail. Free download available from the Eurachem website www.eurachem.org/guides.*

12.3   Random and Systematic Sampling

Key points
— In random sampling all parts of the target have an equal chance of being selected.
— Random sampling may be difficult or impossible to carry out.
— Systematic sampling is often used instead of random sampling.
— It is not possible to make valid inferences about the target unless the whole of it is accessible for sampling.

By definition, an unbiased sampling procedure, under replication, should provide samples with an expectation (a long-term average) of composition equal to that of the target. Individual samples will vary, but the mean of a large number of samples will approach the target composition. This outcome can be ensured only if the sample is a random sample. This implies that every part of the target must have an equal chance of being incorporated in the sample. Random sampling, however, is often impossible, too costly or too time consuming to carry out, and some kind of systematic sampling is used in its stead. The difference between random and systematic sampling is shown in Figs. 12.3.1–12.3.3, using as an example the collection of a composite sample of 20 increments of soil taken from a field. 'Stratified random' sampling is a compromise between the two forms, in which the target is divided into equal parts ('strata'), and each part sampled

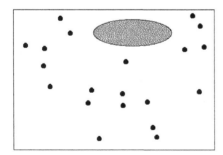   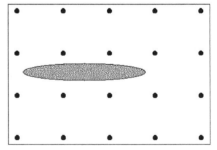

Fig. 12.3.1. Example of random sampling of soil: increments (points) taken from a field. A substantial hotspot (shaded ellipse) could be overlooked in this instance, but a second random sampling would be likely to detect it.

Fig. 12.3.2. Example of systemic sampling of soil from a field: increments (points) taken at the intersections of a rectangular grid. A substantial hotspot (shaded ellipse) could be systematically overlooked.

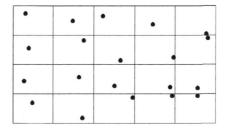

Fig. 12.3.3. Example of stratified random sampling of soil from a field: increments (points) taken at random positions within artificial strata.

randomly. The strata could be real, as when a shipment of coal arrives in a number of railway trucks, or notional, as when a field is divided into equal rectangles (Fig. 12.3.3).

Systematic samples are often as good as random samples in practice, although both are capable of missing an important 'hotspot' (a region of anomalously high concentration of the analyte) in a target. Only increasing the number of increments taken (and therefore the cost of the sampling operation) could reduce this possibility. Sampling procedures that cannot access all parts of the target, however, cannot possibly produce a random sample. Inferences from such samples should therefore be treated with caution, as they may not be based on sound statistical principles.

Notes and further reading

- *Ramsey, M. H. and Thompson, M. (2007). Sampling Uncertainty in the Context of Fitness for Purpose, Accred. Qual. Assur., 12, pp. 503–513.*

12.4   Random Replication of Sampling

Key points
— Sampling protocols do not provide instructions for collecting randomly replicated samples.
— Ideas for several regular sampling scenarios are presented.

Estimating uncertainty from sampling involves the replication of the established sampling procedure. To encompass the potential variation fully, the replication has to be done in a randomised way. Sampling protocols, however, do not provide instructions for how that can be carried out. That may tax the ingenuity of the sampler in some instances, although certain ideas are broadly applicable. If, for example, the protocol demands that increments for a sample be collected at random positions within the target, the increments for the second sample should be collected at new random positions. This is illustrated for stratified random sampling in Fig. 12.4.1. If the target is usually sampled by coning and quartering, it should be re-coned after the first sample is taken and then quartered again. In sampling soil or crops from a field, a common practice is to walk the field in a W-shaped path (Fig. 12.4.2) and collect increments at roughly equally spaced intervals along each leg. To duplicate this adequately, the sampler

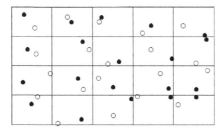

Fig. 12.4.1. Schematic method for duplication of stratified random sampling of soil in a field. Solid circles show the positions of increments for the first sample, while open circles show the positions for the increments for the second.

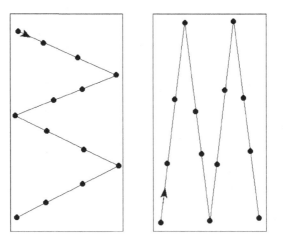

Fig. 12.4.2. Duplicate sampling from a field. Increments (solid circles) are taken at roughly equal intervals along each leg of a 'W'-shaped path.

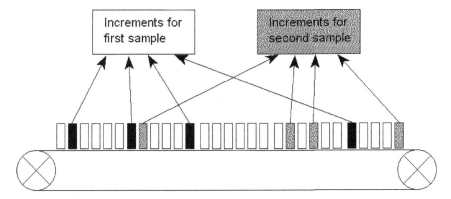

Fig. 12.4.3. Schematic method for duplication of random sampling from a conveyor belt. Increments are taken according to two independent lists of random numbers or times.

could walk a second 'W' in a different orientation. If the test material is presented in a systematic way, on a conveyor belt for example, two different random selections can be made (Fig. 12.4.3).

Notes and further reading

- *Ramsey, M. H. and Thompson, M. (2007). Sampling Uncertainty in the Context of Fitness for Purpose, Accred. Qual. Assur., 12, pp. 503–513.*

12.5 Sampling Bias

Key points
— The existence of sampling bias is denied by some scientists.
— It is difficult to address sampling bias experimentally.
— A useful method for comparing sampling methods is the 'paired sample' approach.

Sampling bias is a contentious subject. Some authorities claim that it does not exist: if the sample is taken 'correctly' (that is, according to the agreed protocol) there is no bias by definition. This is equivalent to saying that the sampling method is akin to an empirical analytical method, where the method defines the measurand. This is a comforting point of view, as the sampler need not worry about bias. Those holding this view are encouraged by the fact that it is difficult to address sampling bias adequately in practice. However, it is easy to see how bias could arise, for instance if the sample was contaminated by the sampling tools or if the sampler misinterpreted the protocol. Bias in sampling could be general (method bias) or specific (sampler bias).

We can consider analogues of methods used to study bias in analytical methods as potential tools for handling sampling bias.

- A certified sampling reference target (analogue of a certified reference material) could in principle address all sources of sampling bias, but would be extremely costly to create, difficult to maintain and could not be distributed to users. Very few examples have been reported.
- Inter-sampler studies with a single sampling protocol (analogue of a collaborative trial) could address between-sampler variation. In these trials, the collection of biases of individual samplers is regarded as a random factor. A small number of such studies have been carried out on an experimental basis, and in some a significant difference between samplers has been found. The 'reproducibility sampling variance' could be used as an extra term in the combined uncertainty. These trials are costly to organise, as all of the samplers have to travel to a number of targets.
- The 'paired samples' approach (analogue of the paired methods approach [§9.10]) is carried out by sampling a (preferably large) series of typical targets by two methods, the method under scrutiny and by an established reference method. All of the samples are then analysed by the same method, so that any analytical bias is cancelled out. The 'paired samples' method is simple to carry out. As a single sampler would normally be involved, any bias detected will reflect the method bias plus an unknown term from the personal bias (if any) of the sampler. The bias between the methods can be characterised statistically by the methods used for comparing two analytical methods (§5.12).

Notes and further reading

- *If sampling bias is ignored, precision alone determines uncertainty and random replication is sufficient to quantify it. Consequently, standard uncertainty and standard deviation are treated almost as identical in what follows.*
- *Ramsey, M. H. and Thompson, M. (2007). Sampling Uncertainty in the Context of Fitness for Purpose, Accred. Qual. Assur., **12**, pp. 503–513.*

12.6 Sampling Precision

Key points
— Replicated sampling and analysis followed by ANOVA is required for estimating sampling standard deviation.
— A multiple target nested design and hierarchical ANOVA is needed for representative results.

Sampling precision can be quantified as a standard deviation σ_s estimated by a replicated experiment. If the act of sampling is replicated in a randomised way, the variation in the composition of the samples obtained is a measure of the precision. However, we have to estimate the composition by analysis and that introduces analytical variation characterised as σ_a. To separate the two sources of variation we have to replicate the measurement as well and use analysis of variance (§4.7). A simple balanced design for this experiment is shown in Fig. 12.6.1, with multiple samples ($n \geq 8$) taken by the same procedure (but randomised, see §12.4) from a single target and with duplicate analysis of each sample. However, this

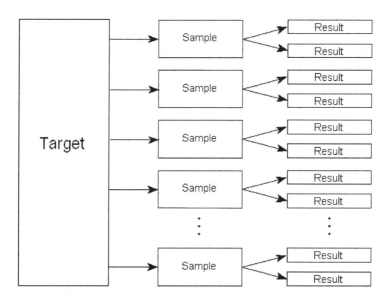

Fig. 12.6.1.   Simple balanced design for estimating sampling standard deviation.

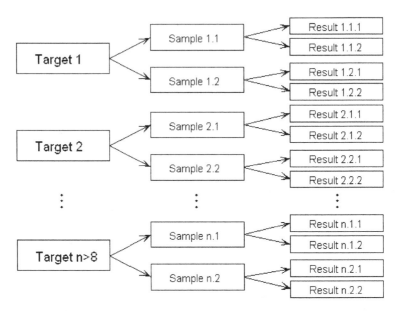

Fig. 12.6.2.   Nested (multiple-target) balanced design for estimating sampling standard deviation.

simple design is based on the assumption that the particular target under study is typical of all targets in the class of material. It disregards the possibility that targets may vary in their degree of heterogeneity and therefore in the value of σ_s.

A greatly preferable estimate, characterising a whole class of material, may be obtained by taking duplicate samples from a succession of different targets of the same type (Fig. 12.6.2). This procedure is also straightforward and involves no extra work, although a greater time span may be required to accumulate the results. Moreover, the procedure points directly to a method for the quality control of sampling (§12.8) in a natural way. The results are treated by hierarchical analysis of variance to obtain the sampling standard deviation. Estimates of the between-target variation and the analytical variation are also obtained.

As an example we can consider the sampling of animal feedstuff and its analysis for aluminium. Twelve successive targets were sampled in duplicate and each sample analysed in duplicate. The results are as shown in the following table and are illustrated in Fig. 12.6.3. In the table

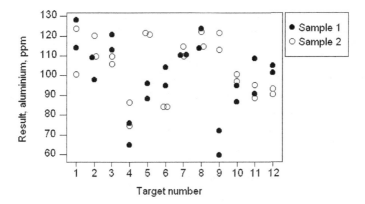

Fig. 12.6.3. Results from a nested duplication exercise to validate the sampling protocol for aluminium in animal feed. There are no indications of (a) anomalous targets or (b) discrepant analytical duplication. Target 9 gave rise to possibly discrepant samples.

the column heading 's1a1' refers to the first result on the first sample, and so on.

Target	s1a1	s1a2	s2a1	s2a2
1	128	114	124	101
2	109	98	120	110
3	113	121	110	106
4	76	65	86	74
5	96	88	121	122
6	95	104	84	84
7	111	110	115	110
8	124	114	122	115
9	59	72	113	122
10	87	95	97	101
11	91	109	88	95
12	102	105	91	93

Hierarchical ANOVA gave the following statistics.

Source of variation	Degrees of freedom	Sum of squares	Mean square	F	p
Between sites	11	7242.25	658.39	1.77	0.170
Between samples	12	4463.00	371.92	7.67	0.000
Analytical	24	1164.00	48.50		
Total	47	12869.25			

From the mean squares we can calculate these estimates:

- The analytical standard deviation component, $\hat{\sigma}_a = 7.0$ ppm;
- The sampling standard deviation component, $\hat{\sigma}_s = 12.7$ ppm;
- The between-target standard deviation component, $\hat{\sigma}_t = 8.5$ ppm. (This statistic is of no importance in the present context except to show that there was little variation in the concentration.)

Notes and further reading

- *The dataset can be found in the file named **Aluminium**.*
- *The above estimates are likely to be rather variable from this (in statistical terms) small experiment. (See §12.7.)*
- *In the example it was reasonable to treat the sampling standard deviation as homoscedastic as there was no indication in the plot of the data (Fig. 12.6.3) of a wide concentration range or a correspondingly wide variation in σ_s. Had such a variation been apparent, some attempt at scaling the data should have been made, for instance by using logtransformation, to render it reasonably close to homoscedastic.*
- *Ramsey, M. H. and Thompson, M. (2007). Sampling Uncertainty in the Context of Fitness for Purpose, Accred. Qual. Assur., **12**, pp. 503–513.*

12.7   Precision of the Estimated Value of σ_s

Key points

— The sampling precision estimated from eight duplicated samples will itself be very variable.

— For best outcomes, an analytical method with standard deviation $\sigma_a < \sigma_s/2$ should be used.

The sampler and analytical chemist must be aware that an estimated sampling precision will be uncomfortably variable. A suggested minimum of $n = 8$ samples is the usual compromise between an acceptable estimate of the sampling precision and the cost of carrying out the experiment, for which $2n$ analyses would be required. If we assume that both errors (sampling and analytical) are normally distributed we can estimate

Table 12.7.1. Confidence limits (95%) for an estimate $\hat{\sigma}_s$ of a true sampling standard deviation of unity (i.e., $\sigma_s = 1$). The calculations are based on an experiment with eight targets sampled in duplicate, and analysed by methods with various analytical standard deviations σ_a.

Analytical standard deviation σ_a	95% confidence limits on $\hat{\sigma}_s$	Proportion of zero estimates
$\sigma_a \ll \sigma_s$	0.5−1.5	0 %
$\sigma_a = \sigma_s/2$	0.35−1.55	1 %
$\sigma_a = \sigma_s$	0.0−1.7	8 %
$\sigma_a = 2\sigma_s$	0.0−2.2	32 %

this variability (Table 12.7.1). With eight replicate samples, analysed in duplicate by using an analytical method of high precision ($\sigma_a \ll \sigma_s$), the relative standard error of the estimated sampling standard deviation $\hat{\sigma}_s$ will be about 25%. Thus the 95% confidence limits will be about $0.5\sigma_s$ and $1.5\sigma_s$. With $\sigma_a > \sigma_s/2$ far worse precisions will be obtained, to the extent that it may be impossible to estimate σ_s. There would be a wide and highly asymmetric distribution of outcomes, with a high proportion of zero results.

The practical rule of thumb is to use if possible an analytical method with a precision $\sigma_a < \sigma_s/2$. Of course, the analyst will not know if this criterion is fulfilled until after the experiment. If σ_s happens to be very small (i.e., the target is close to homogeneous), it may be impossible to estimate its value for lack of a suitable analytical method. In such instances, however, the sampling standard deviation will make a negligible contribution to the combined uncertainty of the measurement and can be safely ignored.

The precision of the estimated sampling standard deviation improves with the number of duplicated samples. Unfortunately, it improves only slowly: in comparison with an eight-sample experiment, 32 samples would be required to reduce the standard error by half. That would usually be impracticable as a one-off method validation. However, if the sampling method is in routine use, data can be collected over many sampling events and the estimate gradually refined. The procedure would be analogous to establishing limits for a control chart while it is in use, as in §10.4 and 10.5. The results would have to be robustified in some way against the possible incidence of atypically heterogeneous targets.

12.8   Quality Control of Sampling

Key point
— A combined sampling/analytical control chart can be constructed from the results of duplicate samples.

We have seen (§12.6) that a 'generally-applicable' estimated value $\hat{\sigma}_s$ can be attached to the sampling standard deviation for a typical target in a defined class. However, the sampler may encounter particular targets, apparently within the defined class, for which the 'general' $\hat{\sigma}_s$ is not appropriate. For such particular targets, the sampling precision may be poor, because the sampling has been carried out ineptly or, more probably, because the target is more heterogeneous than is usual for the type of material. Such instances should be detected if possible, because an incorrect assumption about the heterogeneity will tend to invalidate decisions about the target. Even if the sampling is carried out exactly according to a validated protocol, excessive heterogeneity could make the result unfit for purpose. Quality control of sampling can alleviate this situation.

A simple way of conducting sampling QC is to take duplicate samples A and B at random from each target. Each sample is analysed once, and the mean result $(x_A + x_B)/2$ can be taken as the result for the target. This design is shown in Fig. 12.8.1.

Meanwhile, the difference between the results can be used as an indicator of compliance. The standard deviation of a signed difference $d = x_A - x_B$ for a compliant (in control) outcome would be $\sigma_d = \sqrt{2\left(\sigma_s^2 + \sigma_a^2\right)}$. This value can be used to define control lines for a Shewhart or other control chart so that a single point falling outside the $\pm 3\sigma_d$ limit indicates a system out of control. However, as the order in which the results are obtained is arbitrary, it is preferable to use a one-sided control chart with control lines at zero,

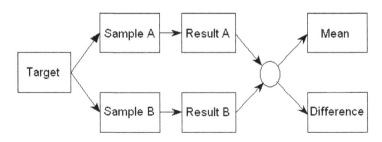

Fig. 12.8.1.   Design for routine sampling quality control.

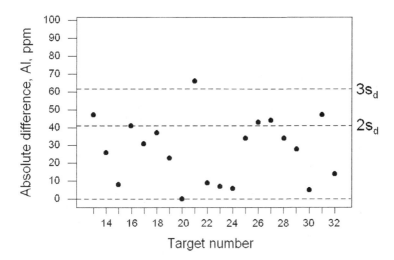

Fig. 12.8.2. Routine internal quality control chart for combined analytical and sampling variation for the determination of aluminium in animal feed.

$2\sigma_d$ and $3\sigma_d$, on which the absolute difference $|x_A - x_B|$ should be plotted. The lines will have the same implications as in an ordinary Shewhart chart.

An example of such a chart is shown in Fig. 12.8.2. The control lines were set according to $\sigma_d = \sqrt{2\left(\hat{\sigma}_s^2 + \hat{\sigma}_a^2\right)} = \sqrt{2\left(12.7^2 + 7.0^2\right)} = 20.5$, using values for $\hat{\sigma}_s$, $\hat{\sigma}_a$ established previously (§12.6). Two out-of-control conditions were detected, at target 21 with a difference outside the action limit, and at target 27 because two successive targets gave differences above the warning limit. It is not clear whether these excursions resulted from an analytical problem or a sampling problem or to a combination of the two. A more elaborate design with duplicate analyses of both samples (as in Fig. 12.6.2) would enable this ambiguity to be clarified, but would obviously cost more to execute as a routine practice.

Notes and further reading

- *The dataset can be found in the file named **Alsamiqc**.*
- *If the concentration of the analyte varies substantially in successive targets, it may be preferable to construct a control chart for relative absolute difference.*
- *An alternative approach to sampling IQC is sometimes applicable, the Split Absolute Difference (SAD) method, which does not require duplicate samples. See: Analyst, 2004, **129**, 359–363.*

Index